Building the Gunpowder Falls -
Montebello Tunnel
1935 – 1940

Ronald Parks

CONTENTS:

INTRODUCTION
A treasure trove and special thanks

Sometime in the late 1980s, renovations began in an old storage section of Plant II at the Montebello Filters on Hillen Road, in Baltimore City. This storage room was to be converted into an office area for the newly hired Water Treatment Engineers. Resident engineers were nothing new to the filtration plant, but somewhere in time, Plant Managers, Maintenance Supervisors and Bureau Chiefs had replaced them. Therefore, the addition of engineers to the staff became a momentous occasion, sparing no expenses to build them a new office and to make them feel welcomed.

It was just by coincidence that the storage area slated for renovation housed a majority of the records from the previous engineers, along with hundreds of glass plate negatives and photographic lanternslides. The maintenance supervisor of that time just saw boxes of junk and gave instructions to throw it all away. Thinking that the information in these boxes looked interesting, I told the laborers working on the project to take all this 'junk' and put in a room on the second floor. This second floor storage area actually did house a bunch of junk and a few years later, when I needed extra storage space, I went over and started cleaning out this room.

While sorting through the boxes, trying to decide what was worth keeping, I decided to keep it all, that I would just go ahead and straighten it out, putting it on shelves. It was while going through the material that I realized what a treasure trove of information was there: old blue prints, engineers' logs, personal journals, water contracts dating back to the early 1900s, deeds to lands obtained by the city through the courts (along with the judge's personal journal, dating back to the 1880s), early photographs and so much more. After I had sorted it all out the best I could, I left it alone for quite a few years.

In 2005, Richard Vann, one of the newly hired engineers, received instructions to put together a history of the water department; mostly just listing all the water contracts and what work was done for each one. However, Richard, being a very thorough individual, started listing everything, from who the mayor was to what the inspector's names were on the jobs. He put this information in chronological order but found that there were gaps in his work, that years were missing from the little bit of records that he had. I then showed him the books and information that I had found years earlier and he started to enter this information into his time line.

In 2006, my boss asked me to assist Richard in what he was doing. We were having electrical problems at the time and I was to work with him to put together a diagram showing all the electrical work done over the years. In the 26 plus years that I have worked here, there has always been a construction project going on, but no one has ever put together an 'as built' drawing of the electrical system. So I set up shop adjacent to Richard's office and via email, he sent me the information that he had. While reading the chronology, I remembered that I had seen additional information, even photos, of things Richard had written about, packed away in the second floor storage area. I decided to go back through all those shelves and boxes of history and see what I could match up to what he had listed.

When I came upon the glass plate negatives, I decided to have some of them processed and turned into photos. This became an expensive proposition so I decided to learn how to do this on my own. My boss gave me permission to buy the equipment I needed, which was no more than a scanner capable of scanning 8" x 10" negatives and Lantern Slides, software to invert the negatives into a positive and a good printer.

While working with one of the lanternslides, I noticed something odd, that in a tunnel, where workers were excavating, there were train tracks that came to a dead end under what looked like a giant boulder. This particular slide came from a box from around 1938, so I asked Richard if he had any information on an event of that year that was of interest. Sure enough, he showed me the Annual Report covering the year 1938 where it was reported that an explosion had occurred in the building of the Gunpowder-Montebello Tunnel.

(Baltimore City Archives Lantern Slide #39)

This notation in the report was only about a half a paragraph long, nothing more than a blurb, so I decided to investigate it further. Searching through all those records that were about to be thrown away 20 years ago, stumbling across filing cabinets that had been stored at the Ashburton Filtration Plant (Home of the Water Engineers in the late 1950s), and researching the archives of the local newspapers, I was able to piece together the story below.

(Note: Upon further research, I found that the picture above was **not** from the tunnel explosion, but rather a progress photo from 1938 of the heading[1] being loaded and wired for

---

[1] Heading or hdgs: a horizontal (or nearly horizontal) passageway in a mine.

blasting. Unfortunately, the tunnel explosion photos are missing from the collection.)

I would like to take this time to thank Richard Vann, Water Systems Engineer, for pointing me in the direction to find this information. His chronology alone could fill a book.

PROLOGUE

Baltimore has a rich history, with it being the *City of First*, *Mob Town*, and the *Monumental City*,[2] and the people and events that helped establish those titles were famous or infamous, depending on your point of view. This is not so much a story of Baltimore's history as it is about the water department and the workers who built the Gunpowder Falls - Montebello Tunnel during the years 1935 – 1940. These workers with names like Watt, Garner, Renbjor, Kavanaugh, and Henthorn all contributed to the monumental task of building a 12-foot diameter tunnel that extended approximately 6-1/2 miles from the Loch Raven Reservoir to the Montebello Filters on Hillen Road.

According to the Annual Report for 1937, submitted to the Mayor and City Council by the Department of Public Works, awarding of the contract for the tunnel work went to the J.F. Shea Company of Los Angeles, at the low bid of $5,389,312.00 on January 27, 1937, with work starting on March 27, 1937. The

---

[2] The Amiable Baltimoreans, by Francis Beirn.

preliminary work, consulting engineers, design, surveying, and purchasing right-of-ways all started in 1935.

The following notes are excerpts taken from the personal journals of the engineers and inspectors assigned to the job, with additional information obtained from the actual, typed and handwritten reports of those mentioned above. Some of these writings were hard to decipher, written in script one day and then print form the next. Many of the notes, memos and journal entries, written in fragmented sentences have been left intact to get a feel for how these individuals wrote their thoughts for the day. By comparing these journals with other documents, such as reports, drawings, time sheets and miscellaneous records, we can determine who wrote what and to what they were referring.

There were four major locations where these men worked. They were: the Montebello Shaft, located at the Montebello Filtration Plant on Hillen Road; the Louise Avenue Shaft, located 11,250 feet from Montebello (just south of Northern Parkway at Louise and Laurelton avenues); the Miller Shaft, located 22,280 feet from Montebello (southwest of Putty Hill and Old Harford); and the portal at Loch Raven Dam near Cromwell Bridge Road (34,650 feet from Montebello).

Today, in Baltimore City and surrounding counties, 1.8 million people[3] still receive approximately 240 million gallons a day from the Loch Raven Reservoir, through a tunnel built seventy years ago. Before the tunnel was built, like any other

---

[3] As of 2007

major city, Baltimore and its water department had its fair share of growing pains to endure before she was able to accommodate her citizens.

Chapter I

Early water history The City grows and burns

Of all the books, pamphlets, journals and web-based information sources available concerning Baltimore's early water history, one of the more comprehensive pieces of work is called: *Baltimore, Its History and Its People,* published in 1912 by the Lewis Historical Publishing Company. A ten-page chapter of this manuscript entitled: *Baltimore Water Works* by Alfred M. Quick, former Engineer in Charge, gives a very detailed account of how Baltimore's water supply was first developed.

Mr. Quick takes us back five years prior to Baltimore's incorporation as a City, when the Maryland Legislature of 1792 passed an act establishing a 'Fire Insurance Company' in Baltimore County. This company was supposed to supply Baltimore Town with water through subscriptions, with the subscribers being the corporation called 'The Baltimore Water Company'. This venture failed and the citizens of Baltimore Town continued to get their water from wells and springs.

Even with Baltimore becoming a 'City' in 1797, it was not until the end of the year 1800 that an Act was passed, enabling the mayor[4] and city council, to introduce water into the city from neighboring springs and streams. This was in large part due to the outbreak of yellow fever in the city. By 1803, there is still no supply of pure and wholesome water to the City, so at the insistence of Mayor Calhoun, the City Council passed an ordinance creating a board of twelve commissioners to obtain water from Carroll's Run, but injunctions from property holders put a stop to this plan.

Then in 1804, property owners took matters into their own hands, forming a joint stock company which chose the Jones Falls as a source of supply, but not until 1806 were they able to purchase a piece of property on Calvert Street for the "works". This consisted of a wheel and pumps that forced water into a reservoir on the southwest corner of Cathedral and Franklin streets. All of the original water pipes and services laid by the water company were made of wood, mostly bored-out hemlock logs, each about eight feet in length. One end of the log was tapered, and on the other end, a bell was hollowed out allowing them to be driven together and then secured with a 2" iron band.

In the Mayor's Message of 1809, which covered the year, 1808, mention was made that the last session of the Legislature of this State passed an act incorporating the Baltimore Water Company. The water company was enacted upon to supply water to the City free of charge for fire fighting purposes only and that the City would be responsible for maintaining and installing the

---

[4] Mayor Calhoun, First mayor of Baltimore.

hydrants and associated piping. In the following year, an ordinance was passed authorizing 'the digging and opening at their pleasure any of the streets, lanes, and alleys of the City, and enjoining the restoring the same forthwith to their formal condition'. By 1811, there were forty fireplugs throughout the city.

The city purchased many of the springs and wells in the next ten years. One great supply of water was located at the corner of German and Eden streets, on the property of Mr. James Sterett. A second supply of water was located on the property of General Stricker. General Stricker had a number of productive springs and authorized the Mayor permission to convey all the water free of any charges by the General to such place as the City desires. The Mayor recommended uniting these other springs, at very little expense, to the spring known as Clopper's Spring.

By 1827, many of the wells in the city had become destitute, with the levels decreasing daily. They would need to be dug deeper. It was recommended that boring machines be bought to dig new wells and that the older wells be filled up. In 1829, the Baltimore Water Company started replacing the wooden mains, which led to the reservoirs, with cast iron pipes. By this time, there were 13 miles of distribution pipe, of which more than half were of wood construction.

On May 11, 1852, the city council authorized the appointment of water commissioners, whose deliberations concluded that, "… the time had arrived when the supply of the city with water ought no longer to be left in the hands of a private corporation ..." and so on December 1, 1852, the Baltimore Water Company was purchased at the sum of $1,350,000. As this money was secured on a loan, additional funds would be needed for the improvement of the "works", so the total loan was increased to $3,000,000. These improvements were completed in 1862 and comprised of the following projects: the impounding reservoir at Swann Lake, now called Lake Roland, a four-mile conduit from Lake Roland to the Hampden Reservoir, and two 30-inch cast iron pipes running from the Hampden Reservoir to the city, at the Mount Royal Reservoir. The Jones Falls supplied all of this water.

The year 1864 brought the realization that the Hampden and Mount Royal reservoirs did not have the capacity to supply the city, so construction on a larger lake began. This lake was at first called Lake Chapman, but later changed to Druid Lake. Built for the capacity of 429 million gallons, it went into operation in 1870. The severe drought of 1872 brought another realization; Baltimore needed to utilize the Gunpowder River or encounter the risk of a water famine in every dry summer. A temporary fix, in 1874, was to construct a dam at Meredith's Ford, on the Gunpowder, where two pumps forced water through a main into the channel of Roland Run, a tributary to Jones Falls, above Lake Roland. This would be the first of many projects concerning the Gunpowder water supply.

A more permanent solution to the City's water problem was to obtain the proper authority, in 1875, to condemn the right of way for the introduction of the supply by natural flow from the Gunpowder River, and in November 1875, the contracts were made for the construction of the entire line. The full scope of work for this project was; an impounding reservoir on the river at Loch Raven along with a dam 800 feet long and 30 feet high. A gatehouse and supply tunnel 12' in diameter and seven miles long, which emptied into a receiving reservoir at Lake Montebello[5]. Another 12' diameter conduit running from Montebello to a gatehouse on the Clifton estate of Johns Hopkins and 40 inch supply mains from that gatehouse to the city. After completion in 1881, for the first time, on September 28, water flowed from the Gunpowder River directly to the city. Note: Because of money constraints, the Lake Clifton portion of this project would not be completed until 1888.

In 1888, Baltimore had expanded its borders from 14.71 square miles to 30.14 square miles, doubling its geographical and political size. This gave an increase of population and water consumption. A new impounding reservoir and pumping station then became imperative. Plans were then made for the Guilford reservoir and the Eastern Pumping Station, with both being completed in 1891. Other works continued through 1899 including construction of a standpipe at West Arlington, a pumping station at Mount Royal, and the dredging of Lake Roland.

---

[5] Property of Samuel Smith. His estate known as "Montebello"

The Great Fire of Baltimore happened in February 1904. The fire raged for two days and more than a hundred fire hydrants were wide open. Water consumption by the citizens of Baltimore was not hampered due to the fire. The Water Department assigned a pumping station to pump directly into the Middle Service main feeding the fire district, and Middle Service mains were opened into Low Service areas to serve other areas where shortages threatened.

Mr. Quick wrote this narrative on the Great Fire of Baltimore and the affect it had on the water system:

> In 1904 occurred the great fire, which destroyed a large part of the commercial and business section of the city. Notwithstanding that many of the mains and large service pipes were broken during the fire, the water supply system was so handled that the pressure and volume of water was maintained and the maximum depletion of any of the reservoirs was not over two and one-half feet. The most important occurrence of this year affecting the water service was the decision of the General Improvement Conference, called together by Mayor McLane immediately after the great fire to consider public improvements, to include in their program a general improvement of the water supply system.

The building of a large, 20-billion gallon storage lake, on the river above the present dam at Loch Raven, was one of two recommended improvements. The other was to abandon the Jones Falls as a water supply all together. The reconstruction of the water service in the Burned District was completed in 1906.

Chapter II

Through the 30s, Consultants, Drought

In early 1910, the consulting engineers, Mr. John R. Freeman and Mr. Frederick P. Stearns, made their report[6] endorsing the plans for improvements at Loch Raven with the addition of building a filtration plant 'for the purpose of clarification and purification of all the water delivered to the city'. This would also include the construction of a high dam on the Gunpowder Falls at a point about one-half mile upstream from Raven's Rock with a new 12-foot tunnel connection to the old tunnel to Lake Montebello. Adoption of this report authorized the water engineer to start plans on the new reservoir at Loch Raven and a filtration plant at Montebello. To prepare for filtration, a small test plant was built at the Loch Raven site to determine the best type of local sand to use in the filtering process. The addition of chloride of lime to the water acted as a disinfectant.

---

[6] Report on the Enlargement and Improvement of the Baltimore Water Supply

Also in 1910 occurred one of the worse droughts ever. There were only 418 million gallons left in reserve, with 1.06 billion gallons used up by the citizens. Because of this, the placing of wooden flashboards on the dam at Loch Raven and waste weirs at lakes Montebello and Clifton helped to increase the storage capacity at each location by three to six feet in depth.

By 1915, both the Loch Raven Dam and Montebello Filters were completed. Lawsuits and the city's inability to obtain the necessary land put a halt to building the dam to the desired height of 280'. The dam's final elevation was 188' with the capability of 192' by the addition of flashboards. The lower elevation meant a redesign of the filtration plant, using the rapid sand filtration method, which is still in use today. Experimentation with disinfection found that less chlorine was needed in the colder months than in the warmer ones.

Once the United States entered World War I in 1917, it was thought best to guard all the vital points of the supply. Hired forces of men, armed with rifles supplied by the Water Department, acted as guards and posts were set at Loch Raven, Lake Clifton and Montebello. The guards reported directly to the Filtration Engineer. The First and Fifth Regiment Infantry took over the guard duties until 1918. This is when the State Militia joined the United States Government service, and the guarding of the vital points on the water supply system returned to City employees. The Police Department then started guarding the City

Water Works, stationing detachments at Loch Raven Dam, Montebello Filters and the Mount Royal Pumping Station.

During the year of 1918, the City once again expanded its borders, increasing its size from 30 to 79 square miles. In 1919, three city engineers[7] recommended raising the elevation of the Loch Raven dam and building a new 12-foot tunnel. In 1920, the consulting firm of Hill and Fuertes were engaged to study the City's water supply problem. Consultants Nicholas S. Hill and James H. Fuertes themselves recommended raising Loch Raven Dam to an elevation of 240 feet, also to purchase the private water companies in the annexed lands. They placed the need for the 240-foot elevation as a top priority, and placed raising the dam to elevation 270 feet as third on the priority list. The Hill and Fuertes Report of 1920 also suggest the construction of the Prettyboy Dam by 1943, in lieu of raising Loch Raven to 270 feet. In 1922, construction on the Loch Raven Dam to elevation 240, along with a balancing reservoir for the increased pressure, was completed.

Through the next ten years, the city built various pumping stations and took others out of service. They filled wells in and closed down public fountains. Typhoid cases, which had decreased by the mid 1920s, would re-emerge by the early 1930s[8].

---

[7] Siems, Lee and Armstrong

[8] The typhoid death rate had risen to 3.3 per 100,000. It was concluded that the prolonged drought was the contributing factor to the rise in typhoid deaths.

In 1932, the City Government again hires consultants to review the status of its water supply. These consultants would form a board of engineers known as *The Advisory Engineers on Water Supply*. The engineers were Messrs. John H. Gregory, Gustav J. Requardt and Abel Wolman[9]. On December 19, 1934, the Advisory Engineers released their report:

1) Immediate construction of a new Gunpowder Falls Montebello Tunnel.

2) Immediately following the completion of the new Gunpowder Falls – Montebello Tunnel, the existing Loch Raven – Montebello Tunnel should be strengthened.

3) Conduct surveys, land purchases, sub-surface explorations and preparation of plans and specifications for the development of an additional water supply should be undertaken at once. (Areas of development looked at by the Board were the Patapsco River; the Little Gunpowder Falls, Winters Run and Deer Creek, and the Susquehanna River).

4) The most favorable additional supply is by the further development of the Gunpowder Falls, by the construction of a new dam at the Big Gunpowder Dam Site, downstream from the Old

---

[9] Abel Wolman, one of the world's most highly respected leaders in the field of sanitary engineering

Loch Raven Dam, to raise the Loch Raven Reservoir to the Elevation 280. Construction of the entire project should be started by no later than 1942. This development will yield 81 million gallons of water daily in addition to the present yield from the Gunpowder Falls.

5) Universal metering is not recommended, but the present policy of the Bureau of Water Supply relative to metering should be continued and sufficient annual appropriations should be provided to assure control of houses and system leakage.

The Advisory Engineers on Water Supply, having completed their work assessment, disbanded on December 31, 1934. Then, on April 24, 1935, these Engineers were retained by the Public Improvement Commission, creating *"The Report to the Public Improvement Commission of the City of Baltimore on Future Sources of Water Supply and Appurtenant Problems"*

Meanwhile, in 1933, the construction of the Prettyboy Reservoir ended, adding about 19.4 billion gallons of water to the storage on the Gunpowder Falls. Storing of the water began on April 10, 1933, reaching the crest level on September 23, 1933.

Chapter III

Preliminary work

On February 4, 1935, the Public Improvement Commission (P.I.C.) sent a memo to Leon Small, Baltimore City Water Engineer, noting in the minutes of their January 28 meeting:

> Resolution was passed that recommendation No. 1 of the Advisory Engineers on Water Supply for the immediate construction of a new Gunpowder Falls-Montebello Tunnel at an estimated cost of $5,300,000 is adopted and this resolution was ordered submitted to the Public Improvement Commission.

Other recommendations considered and the action of the P.I.C.: strengthening existing tunnel, postponed for future study, authority given to Water department to start surveying and preparing plans, resolved and the building of a new dam at elevation 280 ft., postponed. By September 4, 1935, preliminary work had advanced to the point that Mr. Small writes to the P.I.C. asking for direction on the postponed and otherwise unresolved issues brought to the P.I.C. in February. The design of the new tunnel was dependent on whether or not there would be a new dam at elevation 280. Moreover, should the new tunnel be reinforced to withstand a possible future head pressure at elevation 425? This would incur an additional cost of $675,000.00. A motion by Judge Harlan postponed the vote until next months meeting.

The Water Advisory Committee writes to the P.I.C. on October 29, 1935 with this reply concerning the tunnel:

> On the assumption that the next source of water supply for the City of Baltimore would be further development of the Gunpowder Falls by the construction of a new dam with a crest at elevation 280, at a site about three-quarters of a mile north of Harford Road, it is our judgment that the design of the size, character and strength of the proposed Gunpowder Falls-Montebello Tunnel should be predicated upon a delivery of water from a reservoir with flow line at elevation 280 and that provision should not be made at this time for such additional strength as might be required for a greater head which might be ultimately developed at some additional source of water supply.

The P.I.C. resolves, on November 12, all of the above, making it official.

In short, neither the City nor the P.I.C. wanted to spend the extra money on reinforcing a tunnel that had not even been built. Other factors, which may have contributed to the denial of the extra funds, can be found in the personal notes of the P.I.C. during this same period. It appears that the City was more concerned with the building of the Baltimore Municipal Airport than with the building of a new water tunnel. That is a story, all to itself.

The *Engineering News Record* writes an article to the effect that there will be a tunnel built from the Gunpowder River to the Montebello Plant. This starts a stream of inquires from various vendors, trying to sell their goods and services. February 11, 1935, marks the first solicitation from a vendor, received from United States Pipe and Foundry Company. Leon Small writes back that they are still in the preliminary stages and bids have yet to be mailed. Another solicitation comes on February 15 from the Cement Gun Company, expounding on the Gunite Method of lining tunnels. Mr. B.C. Collier, President of the Co. writes that using Gunite after the excavation will greatly decrease the risk of falls and tunnels such as the one at Moffat and Toronto have had great success. In December 1935, Mr. Small invites Mr. Collier to Baltimore to talk about his tunnel lining. Not only do vendors make inquiries, private property owners also start writing. In a handwritten letter, a Mr. Mitchell offers up his property in Glencoe, Md. for the building of a new dam. Mr. Small declines the offer, noting that a site has already been chosen.

Two preliminary estimates for surveys and plans submitted to the P.I.C. on February 18, 1935, were from Mr. Hunt at

16

$74,500.00 and one from J. Strohmeyer at $95,000.00. For whatever reason, Mr. Small sends a memo to the P.I.C. on February 23, requesting $100,000.00 for the preliminary work. They defer voting on this until the P.I.C. meets again, on March 4, 1935 and decides on an allotment of $17,500.00 to cover the expenses associated with the surveys only. All other monies would be voted for on an as needed basis. By May 20, survey work progressed to the point that they now can start on sub-surface work, so Mr. Small makes another request for the remaining $82,500.00 from the P.I.C. As it would happen, the P.I.C. was meeting that very day and they approved the additional funds.

In a response to the news editor of the *Manufacturers Record,* wishing details on the tunnel, Mr. Small reports that bids will probably go out for this project on February 1, 1936. Another file, stored away with this one, was a copy of an article in *Civil Engineering* entitled: *Deep Tunneling for Delivery of Water Supply*[10]. The author, Walter Spear, clarifies the differences between building a tunnel through earth and through solid rock, most importantly, those through rock can be built anywhere without regard to street and property lines. Earthen tunnels, on the other hand, can only go so deep below a water table and must be under existing roads to avoid right of way cost. Deep rock tunnels also produce less shocks from blasting on the surface, there may be however, air concussions coming out of the shafts. These have been known to bust windows but have not caused structural damage in the

---

[10] Civil Engineering, vol. 3, no. 3, 124-125 Walter Spear.

buildings. He concludes his article with "Pressure tunnels in rock may be considered permanent structures." The Gunpowder tunnel will mostly be going through rock.

In preparation to the work beginning, the city recruits a lawyer for consultation on the legal issues that would arise and as we will see later, there will be many. The Department of Law assigns Mr. Wallace Rhynhart to the task, at the request of Mr. Small, noting that Rhynhart was the legal rep during the building of the Montebello-Druid[11] conduit.

One of the first legal issues that needed handling was the obtaining of underground easements. Mr. Small writes to the P.I.C. in October 1935:

> I am sending you attached to this letter a list of the names of owners whose properties the proposed Gunpowder Falls-Montebello Tunnel will cross, and from whom it will be necessary to obtain underground easement for the purpose of construction. The tunnel will vary in depth from 100 to 250 feet, and the easement should be 30 feet wide; for our purpose, it will not be necessary to disturb the surface, as the tunnel will be constructed by drifting[12]. Furthermore, it is not desired to have the right of ingress and egress for

---

[11] Montebello-Druid Conduit connects the clear-water basin at the new Filtration Plant at Montebello with the 60-inch main from Druid Lake to the proposed Vernon Pumping Station.

[12] Drifting - A horizontal passage underground. A drift follows the vein, as distinguished from a crosscut that intersects it. An underground mine in which the entry or access is above water level and generally on the slope of a hill, driven horizontally.

the purpose of repairs. The tunnel will be 12 feet in inside diameter.

Some of the property owners involved in the easements were; Morgan College, Montebello Park Co., a few belonging to Citizens Investment Co., Inter City Land Co., and City and Suburban Realty. The list also included various private owners such as Taylor, Cooper, Hazard, Reynolds, Finger, Hill, Goetze (owned multiple properties in the area known as Grindon Little Farms), Weber, Jenifer and Shanklin. In a final tally of properties and owners, there were 58 owners of 67 properties, of which 14 had been secured gratis. Mr. Strohmeyer estimated that three more would be gratis, with 18 – 22 settling for the fifty-cent offer and 22 - 25 asking for at least a dollar per lineal foot. With one of the properties condemned[13], the total estimated cost for easements came to $22,000.00.

A follow up letter from Small to the P.I.C. in November asks "… how will this legal work be handled?" He suggests that where rights have not been obtained, a temporary easement of five years be negotiated, to expedite future surveying, as the owners are becoming agitated with the constant flow of surveyors on their properties. Attached to Small's memo is a chart showing what has been paid by other cities for similar easements in the constructing of tunnels; The Great Notch Tunnel of New Jersey being the most expensive at $1.45 per linear foot and the least of

---

[13] Condemnation: (law) the act of condemning (as land forfeited for public use)

which was the New York City Tunnel No. 2 at $0.033 per linear foot. The response by the P.I.C. is a memo to the Water Committee, which includes Mayor Jackson, Judge Harlan, B.L. Crozier, S. Cooling and J. Fledderman, asking them to attend a meeting on November 26 to resolve this issue. At the meeting it was duly resolved to recommend to the P.I.C. that:

1. The route of the tunnel is as shown in the blueprints of the engineer.

2. An allotment of $25,000 be set aside for easements.

3. The City Solicitor prepares the Resolution and Documents necessary and informs owners.

At the next meeting of the P.I.C., the above resolutions passed and an Assistant City Solicitor by the name of Alfonso von Wyszecki begins work on preparing the proper forms. Mr. Small reviews and comments on these forms, suggesting changes to certain phrases, such as:

> … it is our thought that where the owner grants us the right to construct a tunnel and we return to him the privilege of using his own land, it will tend to antagonize the grantor. Therefore we suggest that the phraseology be modified that the grantor shall use his own land … in any manner which does not interfere with the tunnel … the City is to be given temporary rights for five years … the Bureau does not desire any surface rights … The instrument provides that the City shall practically acquire the land in fee, and that the

owner is to have merely the privilege of farming the surface. It is our thought that this be reversed, and that the owner retain the land, merely granting to the City a 30-ft. underground easement ...

By January 1936, the acquisition of right-of-ways proceeds and the cost is to be $0.50 per lineal foot. On loan are Employees of the Bureau of Sewerage to begin this task as they have prior experience. In March, one of the property owners requested a flat rate of $1,000 for his 1,805 feet of easement. This would be a $7.25 increase over the $0.50 rate. Mr. Strohmeyer asks Mr. Small to accept this deal. Other property owners follow suit, asking for as much as $15.00 per lineal foot. The P.I.C. passes a motion on March 30 to pay a maximum of eighty cents per lineal foot for easements of the remaining properties and if the deals fell through, condemnation proceedings would begin. Then on May 8, they upped the ante to $1.00.

Two owners, the Intercity Land Co. and a Mr. Scheper, both asked for 'grossly excessive' amounts of money, $25,000.00 and $10,320.00, respectively. The City Solicitor tells Mr. Small to start condemnation proceedings immediately. The P.I.C. votes on the motion to officially authorize this action along with the condemnation of personal properties belonging to Mrs. Charles Class (widow), Christopher Class and wife, Mrs. Jacob Weber (widow), and the Shanklin heirs "... or whomever may be the owner or owners thereof ...". The services of Towson attorney,

H.C. Jenifer were retained, who coincidently is a trustee to the will of a property needed for the Cromwell Portal. To avoid condemnation proceedings, the majority of homeowners agree to settle with the City.

Mr. Strohmeyer met personally with two of the homeowners, the Coleman's, trying to convince them to settle with the City. At their refusal, Strohmeyer remarks:

> He is a peculiar type and insists that he does not want to move, and that we could just as easy move the shaft north to the Horner property. He stopped and would not talk after a while, and referred us to his lawyer. Mrs. Coleman is a similar personality, insisting that she wanted to live on the farm and did not wish to move; unquestionably, if in the condemnation proceedings she puts on the act in court for the jury as she did for us, we will pay a large price. (A settlement of $7,000.00 was finally reached).

By motion set forth at the December 31, 1936 meeting of the P.I.C., a letter of gratitude was sent to Leo McDonagh, Bureau of Sewers, commending his office for their work in securing the right of ways. After all the bickering between lawyers and property owners, an average fee of $0.70 per lineal foot was paid.

Another task in the preliminary phase of this work was for the surveying of the properties to get a centerline for the tunnel, which would lay out the tunnel path and used for easements. In

March 1936, a 56-year-old tree climber named E.J. Hall fell 45 feet out of a tree, knocking him self-unconscious. A call to the Cockeysville Ambulance Company resulted in this response - the contractor was told "… they could not come unless a doctor had been obtained and he stated that it was an ambulance case." So the contractor took the injured man to Union Memorial Hospital in the company vehicle, where he was released ninety minutes later. The doctors reported they could find no broken bones or injuries other than his right elbow and he is "… to remain in bed ten days and take aspirin every few hours for pain, also to apply hot water bag to chest." This was the first record of an injury in the preliminary phase.

As this phase of the work starts to wind down, more information is available for forwarding to prospective contractors concerning tunnel specifications. Leon Small writes to Hitchcock and Tinkler, Engineers and Contractors (Built Moffat Tunnel in Colorado), a list of tabulations of the principal features: Size - 12'-1" inside diameter, Length – 6.8 miles, Number of shafts – 3 each approximately 200 ft. deep, Portals – 1, Type of lining – 2.1 miles steel pipe or pre-cast concrete pipe, 4.7 miles Monolithic Concrete or Gunite, Open cut work – about 2100 ft. of 10 ft. steel pipe to connect to existing dam, Special work at the terminal shaft – consisting of a pump shaft and connections to existing Filtration Plant. Six additional draftsmen were hired for three months, at a cost of $1,950.00 each, to work solely on the tunnel project.

On December 14, 1936, the P.I.C. sends a memo to the 'Honorable Board of Awards' recommending that advertising for bids be sent out, noting that the City has all funds available for this project. Opening the bids took place on Wednesday, January 20, 1937 at 11am, and on January 25, Mr. Small asks the P.I.C. to accept the low bid from the construction firm of J.F. Shea. The P.I.C. accepts this bid on the 27th. The highest bid received was from Angelozzi and Marocco for $8,611,319.70.

Chapter IV

Work begins!

After the Shea Company was awarded the contract, one of the first things that needed to be done was for the City to hire inspectors for this project. As there were no City Service[14] classifications for this type of job, Mr. Small sends off a memo on February 12, 1937 asking the Commission to create the position of "Water Tunnel Inspector". Some of the qualifications required for this position were, High School education, some experience with tunnel work, welding, guniting, concrete work, ability to use survey tools and keep time records, familiarity with P.W.A.[15] and ability to read blueprints.

Although the Annual Report stated that work began on March 27, the first journal entry written was on March 1, 1937. William Watt, an inspector, writes:

Clear. J.F. Shea Inc. Louise Ave. 1 carpenter foreman. 10 carpenters. 5 labors. Unloading material, clearing

---

[14] Precursor to the Civil Service Commission
[15] P.W.A. Public Works Administration.

brush, cutting mudsills, setting plumb post storeroom. John Sullivan – cost accountant. Rulon Bicksterd – material clerk. Clark Ellis – President Building Trades. Roberts – Carpenters delegate. Admas – Carpenters delegate. Busby – Labor delegate, came on job about 2pm ask for Mr. Kavanaugh[16] and left. L.O. Hildebrand, P.W.A.

Starting with this entry, Watt made note of the weather, contractors and tradesmen, and gives a brief description of what the workers were doing.

On March 3rd, George Henthorn, Chief Inspector, makes his first journal entry: "Mr. Hunt[17] phones if I am ready to report on tunnel – advise him yes – says to report at Montebello – going to start field office tomorrow." Upon arriving the next day, no other workers were present, so he reported to the Louise Shaft and met Mr. Kavanaugh and Mr. Gilbert Shea through Mr. Hunt.

On March 4, Watt lists more names and titles of the workers on the job, as well as the vendors where they purchased materials. He also notes the first disagreement between the work groups:

Carpenter foreman – L. Still. Elec. Engr – H.F. Mulherin. Asst Master Mech. – G. Road. Plumber foreman – H. Ackerman. Elect foreman – P. Lambert. W.C. Cole – nails, Lyon Conklin Metal Co., Campbell – sand and gravel. Greenfield Electric Co. Elphinstone

---

[16] Kavanaugh, General Superintendent and foreman for the Shea Co.
[17] John Hunt, Resident Engineer, Montebello Tunnel

Co. – Mixer. Joseph Martin – business agent. Roofers on job raising hell because carpenter put tar paper roof on contractors' office. Referred him to Hildebrandt, P.W.A. (inspector).

For the next few weeks, Watt and Henthorn made entries on the general work progress:

Laborers started excavating trench for sewer drain with intentions to tap into sanitary sewer. Got about 10' dug when Peters arrives. Said it would be cheaper to put a septic tank in so stopped excavation of trench until Mr. Kavanaugh could be consulted. City gang found water supply line was tapped into a gas line. Found water line a few feet away – tapped into it and got water. Ok. Advised Hunt location desired by Kavanaugh for water supply at Montebello. Said he would advise water dept. City gang makes 2" tap in main on Hillen Road for contractors use at shaft. Messer's Small and Strohmeyer[18] out – Mr. Small suggested office be painted a light cream with a dark ivory trim. Also says leave ceiling out of boiler room.

Most of the entries from March 4 until March 30 were in reference to the work done on building the Field Office at Montebello. It took 25 days to complete.

49 carpenters, 2 mixer operators, 1 plumber, and 1 helper, 35 laborers. Carpenters working on main building – setting roof rafters post for traveling crane – roof truss. Setting frame in cesspool. Plumbers setting

---

[18] J. S. Strohmeyer, Baltimore City Bureau Engineer

fixtures in P.W.A. office. C.I.O.[19] sent 14 men on job today. One shovel operator and factory man replacing broken manifold on Lima Shovel. One laborer as watchman. Moved dragline[20] to portal, ready to dig.

On March 17, 1937, Mr. Small wrote to the Shea Co. asking them 'what is the delay' in starting the tunnel. Shea replied on the 27[th:]

> We wish to advise you that we are doing our utmost to get this work underway…special equipment ordered…90 day delivery…contemplated purchasing power from Consolidated Power but were unsatisfactory in making arrangements with them … ordered diesel engines and generators to develop our own power … we discovered a badly depleted heavy timber market around Baltimore and placed our first order for 300,000' BM with Chapman Lumber Co. of Portland Oregon … we expect to open the Miller shaft about April 12 and the Louise shaft on April 19th, the Montebello shaft about April 26th … we have erected a number of temporary buildings … built access roads, and have an average weekly payroll approximating twenty-five hundred dollars, with ninety-nine men at present employed.

---

[19] C.I.O., Congress of Industrial Organizations. Now a part of AFL-CIO.
[20] Dragline – A large excavation machine.

Finally, on March 27, 1937, work proceeded on digging the tunnel. Watt's entry made note: "Excavating portal [at Cromwell] from Sta 346+50 to 346+01.6" [48.4 feet] [21].

Within a month of the job beginning, personnel issues started to come to the forefront of the inspectors' daily writings. Watt entered the following on April 6:

> 2 labors on job without N.R.S.[22] cards. Stopped them from working until time keeper brought their slips. These men were sent out to work while Carter[23] went to the NRS and signed them. Foreman asks me to let them work until their slips were brought out on job. I told him nobody could work with out a slip from NRS.

Henthorn followed up on NRS men with this notation on April 9:

> Louise: George Transfer men have Goo. [Sic] cards and place diesel on foundation, but after conference with Hildebrandt, Kavanaugh and Henthorn, Hildebrandt advises to disregard George Transfer after Kavanaugh says it was a mistake in the contract and will not occur again.

For some reason on April 14, Henthorn starts writing in cursive instead of print. On April 15, he notes a milestone: "Broke ground for shaft at Louise."

---

[21] Sta or Station – Surveying point, one point to another representing distance from first station. In example: Sta 1+12 is 112 feet from starting point. Sta 41+36 is 4,136 feet from starting point (first station). First # left of plus sign (+), is multiplied times 100. Number to right of sign is added to first #.

[22] N.R.S. Possible: National Refugee Service.

[23] Carter: Local Union President.

Through April 18, both inspectors noted the progress that was taking place above ground: "Cinders for road delivered, telephone cables strung, erecting of powerhouse at Louise and change house at Montebello. Framing storehouse and setting diesel engines."

Two more inspectors, William Gardner and E.B. Grantlin, joined the project on April 19, 1937. Also hired, on May 7, was another inspector, N.L. Bayrle. In addition, Robert C. Curtis, a tunnel foreman, was hired. Throughout the years, all the inspectors and foremen would rotate between the different work sites.

The majority of the journal entries for the remainder of 1937 make note of personnel issues and progress. There were a few promotions listed to take effect on April 19: Elmer Johnson and Chris Renbjor, from carpenter to carpenter foremen. Chas. J. Buff, Geo. A. Christensen, James J. Barry and Newman J. Shifflett were promoted from laborers to miners. The problems with employees noted: "Stone not on job. Reported sent home on account of drinking." "George Brown, crane operator reported for work tonight. He worked an hour and due to bleeding from his mouth, relieved by Davidson who worked until 12 midnight. Brown had his tonsils removed and all of his upper teeth drawn 2 days ago and the bleeding was the result of these operations." And progress problems: "Line and grade party gave Mr. Hunt readings on timber arches. It was mentioned that some had sunken below grade." "Heavy storm here before I (Gardner) came on. Current off, was back on when I came to work about 11:20. Burns met me with car at Montebello. Water pump broke due to blast. Men are complaining of head aches due to fumes." In addition, "Children have been throwing fire crackers down at portal at men as they are taking in drill frame to heading."

May 3, 1937, marks the first day of a job-related injury during the construction phase. Gardner notes the incident first, "Mr. Williams reported man hurt on job at 2pm. Fell from ladder, landing on a stake driven in the ground. Injured back. Name Beverly G. Gray." This was followed by Curtis' journal entry, "Beverly R. Gray, Carp. helper fell from ladder at 2pm. He was carrying water up to carps and the ladder slipped. Gray falling, injuring his back. The ladder was not placed correctly. Gray sent to Union Memorial Hospital."

Throughout the next few years, there were numerous injuries including: "Paul M. Northern (miner) working 10pm – 3am shift hurt by flying rock from shot, Clarence Jenkins, Nipper,[24] (Curtis' shift) hands burned by acetylene tank catching fire in tunnel. Burns not bad and, Wachter (powder inspector) struck by motor in tunnel 2:30pm, injuring left ankle. Same being bandaged on job." On June 15, while hauling shaft spoil to the dump on Lakeside Riding Academy grounds, a crane operator, Stallings, dropped the bucket about ten feet with three men in it.

---

[24] Nipper: A tool, such as pliers or pincers, used for squeezing or nipping. One who uses tool.

Joseph Morton, mucker[25], wrenched his right side when the bucket hit bottom and turned over. This was the only injury sustained by any of the three men.

Injuries and treatment within the Montebello plant were handled in a segregated fashion. On July 21, 1937, in a memo from Leon Small to James Armstrong:

> The writer has been directed by B.L. Crozier to add the following name to the list of doctors who shall render medical attention to employees of this Bureau injured in the course of their employment: Dr. R.L. Jackson. Dr. Jackson is colored and to him shall be sent for medical attention, when necessary, all colored employees of this Bureau; this instruction does not affect existing arrangements for the care of injured white employees.

Other personnel matters of the time were included in various memos. Such as a June 8, 1937 letter from Shea Co. to Leon Small, requesting that their employees be allowed to work 40 hours per week instead of the contract specified 130 hours per month. Shea makes note of the fact that, "Some of our good men have become dissatisfied on account of only being permitted to work 130 hours per month and have left, others have made unreasonable demands of their union official, threatening to strike and tie up the work." (This was the first mention of any union involvement on the job and the possibility of trouble with them).

---

[25] Muck: Earth, rocks, or clay excavated in mining. Muckers: To remove muck or dirt from (a mine, for example).

He also notes that New York is starting a tunnel shortly and will take men for a 40-hour week. Another letter followed this one on June 15, stating the same case with an attached memo from Local 273:

> In reply to your urgent and repeated request for competent tunnel workers, I wish to inform you that rock miners and other tunnel workers who follow this work, who transfer here seeking employment, invariably after being informed that this job is a 130 hour per month job, leave, and frankly state, that they would have no part of same and continue on to New York and the Delaware River job where a 40 hour week schedule prevails. The number of experienced hard rock tunnel workers in this section of the country is limited and a great many have asked us to transfer them to other jobs.

In another June 15 letter, Resident Inspector, Louis Hildebrand informs Leon Small that the contractor has violated the specifications of the contract by allowing Norman Heffner to work 132 hours. "This is an un-adjustable, non-compliance and will affect the grant on this project. The contractor should be instructed to refrain from further practices of same."

After a number of correspondences back and forth on this issue, on June 16, 1937, Mr. Small sends off a letter to Abel Wolman asking permission to allow the contractors to work a 40-hour week. On June 22, Wolman responds favorably with this request.

Not only were the rank and file employees dissatisfied with the pay and hours of work, the engineers, who were the bosses of the time, also appeared not too happy with their lot. As documented in an August 4, 1937 letter from Leon Small to the City Service Commission requesting pay upgrades for his engineers, stating this reason – "... have such ratings that their pay and titles are not commensurate with their technical ability, the importance of their duties and their experience." On August 9, the Public Improvement Commission approved this request, pending approval by the Board of Estimates. Then on August 24, 1937, the Commission created the classes with higher pay rates. [Note: This same type of request was made for the supervisors of Montebello in 2004. It took until 2007 to happen!]

On a lighter note, Bayrle offered up this notation:

> December 18 – Vacation. Mr. Kavanaugh gave all city men on the job a very excellent turkey supper at Miller's[26] store at Loch Raven today at 6:30pm. Plenty to eat. Plenty to drink and how – Mt Vernon, White Horse, etc. Those conspicuous by their absence – Bill Iardella, Bill Rogers, G. Henthorn – Chiefy, M. McCabe, Wm Watt, Bill Gardner, John (Pappy) Williams, Wachter, Jack Isaacs. Gil Isaacs and Sam Frazier were there.

---

[26] Miller's Store now called Sanders' Corner.

Chapter V

Union woes

By the end of 1937 and through most of 1938, union troubles seemed to overshadow all other work recorded by the inspectors. Gardner writes this lengthy piece in his personal journal concerning the unions of the time:

December 30, 1937 - A near riot and free for all at portal Dec. 29 – 30, 12 mid. Anderson, shifter. Ross, walking boss. Place, in front of contractors' office, inside of gate. The trouble started yesterday, Wed. Dec 29, during a C.I.O. union meeting between Carter and Richard Ralpud, a miner. It appears that Ralpud came here from the west with a paid up Miners and Smelters Union card and Carter gave him a job as a miner at the portal. From what I can learn, he paid his monthly dues to Carter but refused to pay the required $50.00. When this was brought up at the union meeting, a rough fight taken place between Carter and Ralpud. Tonight, at the above time and place, Carter, with two other cars, total 3 cars with 7 of his followers as body guards came into view just as the shift was coming out of the dry house to go to work at 12 midnight. Carter went up to Ralpud and told him 'he could not go to work and we are all here to fix it so you can't work'. With that remark, Ralpud made a pass at Carter, which was intercepted, by one of Carter's bodyguards, Jim Morgan and a good fight started, lasting 15 minutes between the two men. No one else interfering. Carter yelled to 'shoot the first bastard that moves'. One man, his name I do not know and cannot find out, (but his description you may be able to determine) was a compressor man at Montebello, prizefighter, hair combed back, 5'-8"; 160 lbs. had a revolver in plain view to carry out Carter's orders. Some of the men said they seen more revolvers. Carter had his hands in his pocket. The known men who came with Carter are Armor Burgan, on M. Anderson shift at portal this week, 8am to 3:30. Jim Morgan, mechanic at Louise. The crew of Ralpud shift told me that the revolvers were the only thing that saved Carter and his gang. After the fight was over, Ralpud got into Carter's car. I cannot learn if Ralpud was forced into Carter's car or got into it voluntary. Anyway, Carter came out to do a good job on the man and then the man got into Carter's car does not make sense. Witnesses to the affair are on opposing side.

December 31 – Mr. Iardella[27] was up tonight when I arrived. Was told to get additional reports of last night

---

[27] Wm. L. Iaradella, Assistant engineer.

incident. Richard Smith, chuck tender, on my shift identified Jim Morgan, as the man who fought with Ralpud, also Geo. Morgan, brother, mechanic at Louise, was one of Carter's bodyguards. The man with the gun in view is verified as the one described in yesterdays report. Those who are willing to testify as seeing him with the revolver are Frank Moore and James Semonis. Ralpud lives with James Potter (miner), on one of the back roads near the portal, Royal Oaks – Harford and Joppa Rd. From Potter I obtained this letter, whom Ralpud allowed to bring to work. Also, Carter told Ralpud that there were nine revolvers on the gang when Carter taken Ralpud away. The letter as follows:

White Plains, N.Y.

St. Charles Hotel

Dear Lane:

This will introduce brother Ralpud, whom I wish to be taken care of. You can work him in on the job. He is my responsibility, so the sooner he gets to work, the cheaper for me. If it takes the $50.00 for the Fakirs[28], I will wire you same. Have mailed you, C.I.O. office, in book bunch of pamphlets.

Sincere,

J.R. Carter

---

[28] Fakirs: Beggar.

On December 31, 1937, Mr. Hunt sends a memo to Strohmeyer, Distribution Engineer concerning the fight at the portal:

> [Carter] said the fight at the portal was not premeditated but was only an investigation of the executive committee of Local 273, C.I.O., who had gone to the portal to straighten out certain union matters. He said the fight was started spontaneously, was a fair fight, no guns being in evidence, and that after the fight he took Ralpud to town and had his cuts dressed.

Ralpud concurred with these statements and allowed to return to work.

Union and employee troubles carried over into 1938, and both Anderson and Watt spent considerable time documenting the various incidents. Anderson made this entry on January 3:

> Much controversy going on among the men regarding last weeks riot incident. Also found out that those implicated are plenty scared, having learned that the govt. representatives are investigating who got word to Ralpud to stay in town.

Apparently, they could not have been too afraid of what was happening, as Watt notes on January 4, 1938:

> Crew left job at 7:30pm to attend a protest meeting at Union hall. Meeting was called to vote Carter and his musclemen out of office. The Sunpaper received an anonymous call that there would be a sit down strike at

the portal. A reporter named Kenney came out to the job and asked what time the strike would start. Barns, Garner and Carney told him he was on a wild goose chase.

Along with the internal problems the union was facing, with power and money, problems between the workers themselves came to the forefront as documented in this January 11, 1938 memo from Small to Strohmeyer:

> This man called on me in the office today and complained of the way he was treated on the Gunpowder Falls – Montebello tunnel. He stated he was a miner at Montebello and for reasons he thought were connected with his efforts to improve conditions of other colored miners on the job, he was discharged and expelled from the Union. From my talks with him, he seems to have reached a stage where he is desperate. He appears to be fairly well informed, but deep in his subconscious mine (sic) is the fact that white men generally impose upon the colored people. He left the office with the intention of doing something desperate, and this memorandum is written in connection with this matter. I advised him, before taking any violent action, to talk with the National Association for the Advancement of Colored People.

Anderson continues with his colorful commentary of events through February 1938, making note of the following:

> January 7 – At 1pm I was in contractor's office. O'Conner from Carter's outfit came in the gate and talked to Bergan for about 10 minutes, standing at the south end of tipple[29]. Behind him was a hard looking mug, presumably his bodyguard. Both hard looking yeggs[30]. They left their car outside of gate. I just kept my eye on them, keeping my distance. After awhile they, all three, including Bergan, went outside. In all, they were here ½ hr. nothing unusual happened. January 24 – Werner brought a letter addressed to the attention of Kavanaugh, reprimanding him in reference to the fight and reoccurrence of same on city property, which I think is in order. February 1 – Carter on job 5pm. Don't know what it is about but from all indications and few words here and there, he checked up and stopped the contractor from shifting 65¢ to 75¢ jobs. For instance, the brakeman as chuck tender and the contractor attempted to shift truck drivers in the tunnel. Otherwise, quiet.

March and April of 1938 were quiet months as far as union business went. Most of the journal entries were on the progress of digging the tunnel. However, by mid May, Carter and his gang are back and Watt was documenting their every move. He had this to say of their latest escapades:

---

[29] Tipple: An apparatus for unloading freight cars by tipping them.
[30] Yegg: A thief, especially a burglar or safecracker.

May 11 – Garner, Barns, Gilbert, Renbjor, Doroff and Goetz are ringleaders in movement to get Carter voted out of office. I understand there is about $15,000.00 that can't be accounted for. The above group claim misuse of funds.

May 12 – Carter, Northern, and Burgan on job about 4pm. Carter talked about 1 hour to Barns, Carney and others. [I] heard someone make remark that if he was double-crossed, he would blast someone's guts out.

The next few days of entries by Anderson makes note of a strike and job lockout by Shea. Anderson writes:

May 12, 1938 – A man by the name of Brown, from the C.I.O. in view of a communication from Hubert Gilbert, that Carter threatened to pull him and others off the job for expose of Carter for alleged deficit in his accts to the union. Much controversy taken place between Gilbert, Burns, and myself, we of course assuring the men that we will, positively will, strictly enforce our side according to our specifications. The men who heard the conversation were Lewis Walters, Leonard G? Boyce Nilsson, Britton. P. Also heard that Carter was here on the job on 4 to 12 shift. I told Ross there should be a safety miner on this end.

May 13 – Job shut down 12 midnight. Orders from C.J. Kavanaugh. Trouble among the men about union policies is the cause for the shut down.

In a separate document, Shea notes in a May 26, 1938 memo to Leon Small, advising him of a strike by the local union:

We wish to advise you that on May 13 about midnight, all the workmen, whom the J.F. Shea had employed ... were ordered to stop work by representatives of the local C.I.O. Union. These representatives informed us that due to an internal friction between factions of their union, and impending trouble, it was necessary for them to temporarily stop our work, in order to bring about a settlement of their differences. This stoppage of our work occurred through no fault of the J.F. Shea Co. It was on May 24 that we could resume our work again ... in view of these facts, we be granted an extension of time commensurate with the time lost.

Anderson continues with his journal entry:

May 14 – When Burns and I came on at 11:30 Friday night, there was posted in the washroom a notice headed 'Miners and Smelters Union' but unsigned other than type written 'Strike Committee'. The body of the notice was to the effect that a meeting was held at hall 2pm Sun. Gustafson shift was in the tunnel. His men knew nothing about the notice. All the crew on Gus. shift got together. Hell was up. They all agreed to go after Carter. About 12:30pm, Brown, the President of the union, came in, tore down the strike committee notices, and placed another in its place, signed by Brown. The body of the notice was to call a regular meeting sat 2pm at Bohemian Hall, Preston St. At the same time this gang found out that Carter was out after some ones hide, accompanied with a cop in uniform, later a gang of henchman. This caused their ire and then gang went after Carter. Where they will meet, I don't know. The men want Carter out.

May 16 – Shut down acct. strike. The only men working are the three diesel operators. Men want to go to work. Kavanaugh holding them up for what reason no one knows.

The strike lasted through May 23, 1938 and Watt made most of the union related entries and exerted his authority:

May 16 – Job shut down. Carter voted out of office as business agent of Local 272 C.I.O. May 17 – Job shut down. No arrangement between C.I.O. and Kavanaugh yet.

May 18 – Job shut down. Ross told me that Kavanaugh went to New York today to meet one of the Shea's [J.F. Shea Company]. Ross also told me that the union drew up a new agreement and it was not acceptable to Kavanaugh.

May 19 – Job shut down. M. Anderson told me tonight that Carter called a union meeting and told the men he would get them 8 hours a day, 5 days a week. The men refused Carter's offer. Also understood Kavanaugh would order work resumed 'if' Carter remained as business agent. Kavanaugh attended the above meeting but was in a side room, out of sight. Several of the men on job have openly stated that they believe there is a financial agreement between the contractors and Carter or between Kavanaugh and Carter.

May 23 – Reported around job this morning that Carter had a restraining order against Kavanaugh to stop him from starting job.

June 3 – Barns, Robertson, Doroff, Gilbert bros, Graner, and others on job tonight discussing union affairs.

June 4 – Three men from Building and Construction Trades Dept., A.F. of L.[31], passing hand bills about a meeting June 5 Lehman's Hall, 2pm. Robertson, C.I.O., Hall A.F. of L. to get the hell off the job and stay off.

June 6 – Hall and 3 other A.F. of L. men on job this morning about 2am pasting handbills about a meeting tonight 1222 St. Paul St. A fight almost started between Hall, A.F. of L. and Barns, muck machine operator. I ordered Hall and the men with him off City property. When I escorted Hall and the men with him to the gate, he told me if these fellows wanted trouble, he had a riot squad of his own. Remarks were made that these men had guns but I did not see any. Hall asked who I was. Told him I was Bill Watt, worked for J.J. Hunt, and was employed by Bureau of Water Supply. Told Barns if organization to which he belonged had any more union affairs to discuss to do so at their hall and not at the Portal, as was the case last week on two different nights.

Bayrle made this union notation on June 6, 1938: "A.F. of L. stuck posters everywhere in the yard and outside on fence.

---

[31] A.F.L.- American Federation of Labor

Order(s) are now for city inspectors to keep all people outside of fence except the men working here and those who have permission to enter from Mr. Kavanaugh."

Troubles with the Unions became bad enough that on June 17, 1938, Mr. Small wrote to Police Commissioner Wm. P. Lawson:

> As the matter affects the peace and quiet of the community, I am attaching to this letter, for your information, a report of our Water Tunnel Inspector, G. G. Tyrell, regarding a disturbance that occurred on the morning of June 15 at the Louise Shaft ... As similar disturbances have occurred previously, we request you to instruct your forces to visit these various construction shafts at frequent intervals; two rival factions are striving for control of the working personnel on the tunnel ... It is our fear that the disturbances occurring among the workmen on the tunnel will eventually result in a fatality, and there is a possibility that the visits of the police cruiser at frequent intervals may tend to prevent such an occurrence.

On June 20, the Commissioner replies that he will give the matter special attention. Mr. Small follows this up the next day in a memo to Shea, requesting that they fire employees "who are objectionable to the City authorities."

Chapter VI

Premature blast

The worst of the job related accidents happened on July 20, 1938. Bayrle wrote in his journal, "Premature explosion occurred at 6:30am while the heading was being loaded. Ten men killed and eight men injured. This is the worst accident and sight that I <u>ever</u> have seen or <u>hope</u> to see again." Watt made this entry for that day:

> In tunnel at Montebello this morning about 7:20am. Watched the removal of men killed by premature explosion in heading. Mr. Kavanaugh told me the top left cut hole was where the explosion happened. I believe some of the dynamite on the Jumbo[32] also exploded. Out of tunnel about 8:45am. All bodies removed.

The annual report for the Department of Public Works had this small notation by Associate Engineer Strohmeyer concerning the tunnel explosion:

---

[32] Jumbo: a multi-drill machine, which can drill more than one drill hole at a time. Normally used for drift development because it can be repositioned easily for blasting and mucking.

46

The progress on the job, which is about 60% complete, has been satisfactory to date. A labor strike delayed the work for nine days in May. The work in the Montebello heading was delayed about two weeks because of an unexplained premature blast, which occurred on July 20, about 6:30am, while the heading was being loaded. Ten men were killed and six were injured.

The local papers published these reports concerning the tunnel accident:

> _Baltimore American Newspaper_, July 21, 1938: "Federal investigators spend 2-1/2 hours probing fatal blast ... examine earth and unexploded sticks of dynamite, personal property of miners taken to Northeastern District police station ... State's Attorney's office might join in the inquiry ... Mayor Jackson, has acting chief engineer join investigation ... survivors quizzed ... explosion 145' below surface ... nine men killed instantly, one died in hospital, four injured."

> _The Washington Post_, July 21, 1938: "18 men, 200' below surface all killed or injured. Caused by premature explosion of 450 pounds of dynamite." (This caption was under a photo of the men as they were being taken out of the tunnel to awaiting ambulances.)

> _The Washington Post_, July 21, 1938: "Baltimore, July 20 - ... dynamite blast killed ten men and injured

six others ... 450 pounds... killed seven of the victims instantly ... all were colored."

The Washington Post, Aug. 6, 1938: "Baltimore, Aug. 5 – Officials investigating discovery of two sticks of dynamite in the Montebello water tunnel received a report today one of the sticks may have been made up somewhere else and taken to the shaft ... police increased their guard ... two coroners ruled their deaths were accidental."

One month prior to the tunnel explosion, Watt had made this entry: "June 23 – Kavanaugh and 2 detectives on job about 5pm. Looked around diesel house. Inspected powder houses and talked to Lambert." There was no explanation as to why this entry was made or what they were looking for and then on June 24, Watt states, "Carney released by police. Here at 7:30pm." Again, no previous entries were found as to why Carney was taken into police custody.

After the tunnel explosion of July 20, 1938, there are many journal entries, memos and reports on the investigation and some remarks on speculation. On the following day, Frank Bender, a member of the Committee for Industrial Organization (C.I.O.) sends this memo to Mayor Jackson:

Honorable Mayor:

I am addressing this letter to you relative to the explosion that occurred yesterday morning in the Montebello Tunnel, taking the lives of ten of our members in the International Union of Mine, Mill and Smelter Workers. The damage inflicted by the explosion has left sorrow and suffering in the many homes of those caught in the blast; this cannot be changed, we can however, make every effort to find the cause of the explosion, if it is possible to do so.

With this thought in mind, I hereby offer my services with more than thirty years experience as a coal miner, to help those making the investigation to find the cause of the explosion. I also recommend Mr. John Carney, President of the Union that has jurisdiction over the union's affairs on the tunnel project; who is an electrician, and was so formerly employed in this tunnel.

Both of us shall be glad to give every aid possible to determine the cause of the accident, if it is desired.

The Mayor then forwards the above letter to Chief Engineer Crozier on July 22, 1938, stating "… would be very glad to have you accept Mr. Bender's offer, if possible." Crozier reroutes the letter to Small who replies to Acting Chief Engineer, Frank Duncan, suggesting that Bender get in touch with the Shea Company, "… any further activities in connection with the investigation of the explosion should originate with the contractor …." In this same letter, Small list those already doing an investigation – Maryland State Board of Mines, the State Industrial Accident Committee, the Federal Bureau of Mines, technical advisors of the Hercules Powder Company, members of the contractors organization, the State's Attorneys Office, the City Law Department and this Bureau.

For the journal entries concerning the tunnel explosion, Mr. Watt only had three entries, but Mr. Bayrle had a lot to say.

Watt writes:

July 26 – Attended safety meeting at Millers store. Meeting called by Haller, State Accident Commission. It was an open meeting. Several men expressed ideas as to the cause of the explosion at Montebello. No definite solution as to the cause given.

August 2 – Found stick of dynamite in sand box of motor. Also, found stone laid against blasting cap in primer at Montebello. Isaac went to Montebello about 8pm.

August 3 – Detectives on job at Montebello and Louise.

Bayrle writes:

July 21 – E. Lehnert, explosive inspector, sent out to work in Lesser's place. He insisted on going in the tunnel tonight with the diesel man, to look at the heading. I did not go in. No work in tunnel until after coroner's inquest. Mucker acting as watchman.

July 25 – Mr. Bruno, electrical expert from US Bureau of Mines, Mr. Hunt, Mr. Kavanaugh, Mr. Lambert, Mr. Rood, Mr. Glaze, Lehnert and myself were in tunnel heading today. Mr. Bruno was testing all electrical appliances and hook ups for possible leaks and short circuits. Every test was negative. Mr.(s) Diggs, Hildebrandt and 3 P.W.A. investigators in tunnel today, checking up. P.W.A. investigators were Mr. King – F.B.I., L.J. Graham, and A. Topoleski. The smell of burned flesh and blood etc in the tunnel is sickening.

July 26 – No work in tunnel yet but permission was given to clean up and do whatever repair work necessary in heading. Mechanics crew went in tunnel today and started cleaning up and repairing "Jumbo" frame, etc. at 10:45am. Mr. Nice, of Hercules Powder, also in tunnel today checking. Mr. Kavanaugh had a photographer take pictures of heading before and after cleaning operations. Hunt, Schudel and I in tunnel taking pictures of Jumbo and heading today.

July 27 – Station heading before shot – 52+56.7. After shot – 52+65.4. Advanced 8.71. This is the first shot made since premature explosion on July 20. (It took 1-1/2 hrs to load). Mr. Kavanaugh, Mr. Dailey, and Mr. Nice were in the heading when it was loaded. Twenty-two primers and approximately twenty-two sticks of powder were taken out of face – placed there on 20th before premature explosion occurred. The smell is still terrible in the tunnel.

August 2 – Phillip Doroff and Roy Graner in yard tonight at 10pm. T. Taylor told them to get out and stay out. Today, on the day shift, a stick of powder (60%) was found jammed behind the cables running from the wet cell batteries to the control box on one of the motors. This powder was put there deliberately and it could not have fallen where it was. It was not the motor that was in the heading at the time of the premature explosion on July 20. Someone is trying very hard to either scare the Negroes off or else kill somebody. Tonight Burns went over the primers in the powder house and he found one with a stone, the size of a quarter, pushed inside the powder and against the business end of the cap. This was evidently planted in McFadden's primers. There was also another with the end of the powder hollowed out, so that if it had been used, the bear cap would have struck the rock surface first. There were also four primers without shunts on them and the wires not twisted together. As there is no work at the portal for the powder inspectors, they will double up at Montebello starting tonight. I brought Isaac down here.

August 3 – Lehnert and Isaac on duty at shaft. They are going over primers and powder very carefully before taking it out of the powder house. The chief engineer has asked the police to assign men at Montebello and Louise shaft on each shift until further notice, starting today at 4pm. Detective Menendez on duty 4pm until 8pm, Detective Madigan on duty 8pm until 12m.

August 4 – Mr. Sharretts, Assistant States Attorney and a P.W.A. investigator with a stenographer questioning men 4pm-7pm. Shift was delayed 1-1/2 hrs due to investigators questioning T. Evans, mucker operator, from 4pm – 5:30pm. Carney, union president, and Sinnes, secretary-treasure in yard tonight at 10:45pm posting notices for meeting on Saturday, the sixth. Menendez was going to arrest them but Kavanagh interceded for them.

August 5 – Investigators here today were questioning Sullivan, powder inspector. Dailey fired Pritchett today.

August 15 – This morning, Sullivan could not account for eight primers he never counted them ahead of time and I think everything ok. Stromberg in yard today to see McCullough Electric. I told him to get outside the gate and I called Mac for him.

August 16 – Drew sketch for Dailey, walking boss today, showing positions of Jumbo frame, jack car, electric locomotive, and powder car, before and after explosions on 20th. August 18 – Mr. Creekmore, staff artist of the Sunpaper in tunnel today, 9:45am to 2:30pm, was drawing black and white sketches of the different operations, by Mr. Hunt's permission.

Bayrle adds this personal note on August 20: "Aunt Stella buried today 8:30am. I was pallbearer. I did not get on job until 11am." He then continues with the business of the explosion:

Mr. King of the F.B.I. and Mr. Topoleski of the P.W.A. questioned me today from 1pm to 3pm about the explosion on July 20 and about primer that Burns found with the stone in it on August 2. A stick of powder was found on motor the same day (Aug. 2). I was questioned in room 437 of the Emerson Hotel and a stenographer took down all conversation. I was told the testimony would be typed and I would be asked to sign it.

Chapter VII

Reports on the investigations

In addition to the daily journals kept by the inspectors, there were numerous reports made, such as the following ones, stapled together and submitted to Frank Duncan, Chief Engineer, from Leon Small, Water Engineer. The first report being from inspector Henthorn to Mr. Hunt on the day of the accident:

Mr. J. Hunt,

July 20, 1938

Construction Engineer

Dear Sir,

Confirming my verbal report to you and Mr. Strohmeyer, would advise that upon Hamilton picking Burns and myself up at 39th and Greenmount Ave. this morning, at 7 o'clock, he advised there had been an explosion at Montebello in which Lesser and several other men had been injured or killed and that Bayrle had taken Lesser to the hospital.

Deciding that someone should be at Montebello, I told Hamilton to takes us there and phone the gang going off at the portal to come to

Montebello in the 314, from where they would be taken to the car line. I asked Hamilton if you had been notified and he said yes. Accordingly, we arrived at Montebello at 7:05am and went direct to the shaft. There we found the government inspector Diggs, the first aid doctor from Louise, the track foreman Pritchett and several workmen on the shaft platform and three fire department ambulances backed up against the platform.

We made a quick survey of the surface and saw the shift foreman Reed, the electrician Elliott and the motorman Geary sitting on the running board of an automobile with their faces covered with blood and their clothing torn into rags. Reed was the worst. I asked them all about their injuries. The first aid doctor came forward and said he had examined them and they were not seriously hurt. I then told them all to go to the change house and wash up and change their clothing. Reed did not want to go, stating there were some dead men in the hole. I told him there were plenty here to take care of them. [I] persuaded him to go. We then returned to the shaft platform and talked to Pritchett. He advised the explosion occurred about 6:35am and that the walking boss Daily and a couple of men were now down in the tunnel but no one could get up to the heading on account of the smoke. I asked him if the air was on and he said yes. Carpenter foreman Still came along and we decided the heading should be clear as it was then 7:15. We looked across the yard and saw Mr. Kavanaugh coming on a run. Master mechanic Ruud was there, so we all; Kavanaugh, Ruud, Still, Burns, myself and several workmen got on the cage and went down the shaft.

Upon reaching the bottom, everybody started on a run for the heading. About halfway in we encountered a dense bank of powder smoke, in which we could not see anything except the flare from flashlights some were carrying. We had not gotten very far into it when we heard the motor coming down from the heading. As it passed us, it looked like two men lying on it, whether they were dead or alive I do not know, for they did not stop. After they passed, we continued to grope our way through the smoke, breathing through our handkerchiefs. When we were about half way through the smoke, which we learned afterwards, we heard the motor returning from the shaft. We got under a light and waited for it. When it was near enough we hailed it and found they had a flat car. The day shift boss Witzke and several of the day gang with them. We got on and went to the heading.

The bank of smoke was about 350' long, but when we got to the heading the air was perfectly clear and all started in to gather up the dead. All attention was centered on them.

#2

The first dead man I saw was lying on the floor, about 20' in front of the jumbo. I kept walking towards the heading and saw a motor and the jack car about two or three feet ahead of the jumbo. I got up on the jack car and by the aid of a flashlight, I borrowed, I think from Mr. Still, I made a general survey. One man was lying between the jumbo and the west wall, another was laying face down between the jumbo and the east wall, another body of a man without head or legs was up in the east corner of the head. It was very dark up in there, the only light we had were the several hand flashlights. Daily, the muck machine operator Evans and the workmen were bringing men out from between the heading and the jumbo and along the west side of the jumbo. We could not use the east side on account of the man laying there. Mr. Kavanaugh and Daily were directing and helping to remove the men. They found that the one on the east side of the jumbo was pinned down by the jumbo, so Mr. Kavanaugh called for a jack. It was some time before the jack got there but when it did, they jacked the jumbo up and got the man out. During this time we noticed sticks of dynamite and drill steel was scattered all over the floor. I told Burns to look out for the dynamite and when he got a chance to examine the heading. There was not any fresh muck in there.

Everybody, except those handling the dead, began gathering up the dynamite and the steel. I kept calling to handle the steel carefully to keep from creating any sparks. Burns was looking after the dynamite. At first, they thought they would move the jack car up under the jumbo and take the jumbo out of the way, so several got on their knees with flashlights to see that no dynamite was under it or the motor but Mr. Kavanagh decided not to move it. We had gathered up a five-gallon can and two boxes of dynamite up to this time.

Burns had gotten a chance to examine the heading at this time and reported to me he had counted sixteen holes had dynamite in them and that one cut hole had exploded. I went to the heading with him and found the cut hole was about 4 or 5" in diameter and was located on the west side of the vertical centerline and a little distance below the horizontal centerline. Mr. Daily was standing along side of me and I asked him if he thought the one hole exploding had caused all this damage and he said yes and I asked how. He replied by concussion. It seemed strange so many men could be torn to pieces by concussion from the one hole.

At this time all the dead had been gathered up and were out in front waiting for the motor to take them to the shaft, so I told Burns I was going out but for him to remain until I sent Keim down to relieve him then for him to come to the top. Accordingly, I waited for the motor to come in and when it arrived, I found Mr. Still had gone out for floodlights and brought them in. On the flat car with the motor was Hildebrand, Diggs, Peregoy, Still and several more men. They unloaded and loaded the body of the one man and another and I went out with them. The body of the one was wrapped up in a raincoat.

Upon arriving on the top, I came to the office and made my report to you, then sent Keim down to relieve Burns. Burns came up with the dynamite he had recovered and had it in the steel box magazines. He reported he had recovered five boxes of dynamite and twenty-one primers.

We left Montebello about 9:30am for the portal.

(Signed) George W. Henthorn

The next report on the incident was from the explosives inspector, Burns to Mr. Hunt:

Mr. Hunt, Construction Engr.

Baltimore, Maryland

Loch Raven – Montebello Tunnel

July 20, 1938

Dear Sir:

When Mr. Henthorn and myself were picked up this morning at Greenmount Ave. at 7:00am, we were told by Mr. Hamilton that an explosion had just taken place at the Montebello Shaft. He said that Mr. Bayrle (Water Tunnel Inspector) had taken Mr. Lesser (Explosive Inspector) to the hospital. We proceeded to go at once to Montebello arriving there at 7:05am.

Under instructions from Mr. Henthorn, I went at once into the shaft. Messrs. Kavanagh, Still, Rood, Henthorn, 2 workmen and myself went in together. We were in the tunnel face about by 7:25am. After the dead were all removed, I went to the heading and asked Mr. Henthorn make notes of the following:

There were dynamite in 16 holes. One top right hole had a primer hanging from it by the wires.

It looked to me like one (1) cut hole and exploded, it was the one just below the spring line on the left. I put a safety flash light to this hole and it looked like it had gone off about two (2) feet from the collar, the back of the hole looked like it had not been touched.

A tamping stick was still in the cut hole over this hole that had exploded.

The two magazines were on the floor of the tunnel, one about thirty-feet from the face and the other about forty-feet from the face, both of the magazines were badly damaged.

The left front of the jumbo looked like it was damaged, as was also the platform.

There was drill steel strewn all about. We gathered up off the floor of the tunnel five cases of loose dynamite and twenty-one primers, these primers still had the shunts on.

One dead man held a primer in his hand.

I suggested to Mr. Kavanaugh that he get the magazine truck and take this to the main magazine, to be picked over by his powder men, this he done.

At 9:20am Messrs. Sam Frazier, Keim and Leonard came to the tunnel. I then left the rest of the dynamite lying around for them to get. I came to the top and brought the five (5) cases of dynamite and twenty-one (21) primers up with me.

Mr. Henthorn and myself then proceeded to our destination, (Loch Raven Portal) at 9:30am.

Respectfully submitted

(S) Howard F. Burns

Explosive Inspector

The next letter to Hunt was from inspector Bayrle:

Wednesday, July 20, 1938

Montebello Shaft (North Heading)

12:01am to 8am

Mr. John J. Hunt

Resident Engineer

Loch Raven – Montebello Tunnel

This morning at 6:30am, Thomas Evans, muck machine operator on this shift, came running up to the City office and said there had been a premature explosion in the heading. He said he knew this because he heard it go off as he was nearing the shaft on one of the motors, which was bringing the steel car back to the cage. He sent the motor back into the heading and he came on top to get help. Evans and the motorman were uninjured. Evans told me that the powder car had just gone into the heading before the steel car came out, and that there could not have been many holes loaded at the time. He stressed the point that there was only <u>one</u> report[33] heard.

I immediately called the Louise Ave. office and told Phillip Dailey, the day walker, to come right over to Montebello, and then I called for all the ambulances that the fire department could spare. It was then that I called you at your home. Dailey arrived as I was talking to you on the phone, and, he and Evans went right down into the shaft with one of the top men, and I was not able to talk to him. In the mean time the cage came up and on it were Phillip Lesser, Powder Inspector, Charles Elliott, Electrician, Chris Tiggle, Chuck Tender, and Walt Colston, Chuck Tender. These four men were shocked, cut and burned pretty badly. Tiggle being the most seriously hurt, was sent to the South Baltimore Hospital in Dailey's car. One ambulance arrived by this time and I had one of the firemen call for more of them.

Mr. Pennington, First Aid Man arrived from Louise Ave. and rendered aid to the above named men. I then took Lesser to the Union Memorial Hospital for treatment, in one of the City cars.

---

[33] Report - a loud noise, as from an explosion: *the report of a distant cannon*

The injured men named above were able to get out of the heading on one of the motors. They said when they left the heading the smoke and gas was terrible, the lights were out, and they could not hear or see the other men.

The cause of this premature explosion is something that I cannot point out definitely as there are several things that could have happened, even though our inspection regulations <u>were</u> carried out to the letter.

We know now that there was only one thirteen foot hole cut hole on the west side of the face that exploded.

As I understand it, the exploded hole was either loaded, or in the process of being loaded by a miner named Charles Durham (known as "Spider") and a chuck tender named James Hough. These two men always drilled and loaded the holes on left bottom machine.

You already have a statement from Lesser, so there is no need for me to repeat his version of the accident.

The last report I have lists ten men dead and six injured. The names of these men I think you have.

2.

Tonight at 9pm I went to the South Baltimore Hospital and talked to John Reed, the shift foreman, he also says he only heard one report.

I tried to see Chris Tiggle at the same hospital but it was too late and the nurse said he was sleeping. An interview with Tiggle may give some additional information.

Respectfully yours,

Norman L. Bayrle, Inspector

The following reports were typed on the same page, unsigned and taken at South Baltimore Hospital on July 21, 1938:

John Reed, Foreman:

Reed had checked primers men had at front of Jumbo, had taken extra primers back to Lesser, who was at back end of jumbo at flatcar where primers were sorted. Had then gone toward shaft to check airlines or something. Started back to heading had passed powder car and had started to crawl thru center of jumbo when shot went off. Had left side of face toward heading as he was crawling from left to right. Did not see flash. After blast could not see because of smoke was groping around tunnel near face when found by Walker Dailey and sent out of tunnel.

Reed took Bureau of Mines Safety course in California.

Chris Tiggle, Chuck Tender:

Had been back from face checking bull hose and had started back to face. Was at powder car when shot went off. Thought powder car fell on him. Helped by Lesser to get back to California switch where tunnel was clear and came out on motor with Lesser, Elliott, electrician, Colston, Chuck Tender, and Geary, motorman.

John Hunt takes all of the above reports and does a summary, sending them to Mr. Strohmeyer:

Mr. J.S. Strohmeyer:

July 22, 1938

At about 6:30am 7/20/38 a premature explosion took place in the Montebello heading while it was being loaded, causing the death of 10 men and injuring 6 more. Since then there has been a number of visitors and investigators at the site of the Montebello shaft. A list of these persons follows:

July 20th, 1938

Mr. Small, Water Engineer

Mr. Strohmeyer, Distribution Engr.

Mr. Fledderman   } P.I.C.

Mr. Ralph Sharretts   } P.I.C.

Mr. Limback   } P.I.C.

Mr. Limback's Asst.   } P.I.C.

Mr. Marshall   } City Law Department

Mr. Henkel   } City Law Department

Mr. Roth   } City Law Department

Mr. Lefko   } States Attorneys office

Mr. Douglas Sharretts } States Attorneys office

Mr. Carr   } State

Mr. Green   } Accident

Mr. Hallor } Commission

Mr. Bill Ellis} Hercules Powder Co.

Mr. M. A. Nice} Hercules Powder Co.

Also numerous police and fire department officials.

July 21, 1938

Mr. J.J. Rutledge } Md. Bureau of Mines

Mr. Frank T. Powers} Md. Bureau of Mines

Mr. C.J. Rowe   } Md. Bureau of Mines

Mr. J.E. Tiffany } U.S. Bureau of Mines

Mr. R.D. Currie } U.S. Bureau of Mines

Mr. W.F. Fene   } U.S. Bureau of Mines

Mr. Graham – Division of Investigation, P.W.A.

The U.S. and Md. Bureau of Mines men made an inspection of the heading and also visited the main powder magazine on the Shanklin property to look at the recovered dynamite and exploders, the South Baltimore General Hospital to talk to John Read, shift foreman and Chris Tiggle who were injured in blast. I attended a meeting at 22 Light St. room 606 9:30am to 12:15pm 7/22/38, at which time I answered questions asked by six investigators.

We recovered about seven cases of dynamite and at least 21 primers (more primers still in holes in heading). There were eight cases of powder taken into the heading. The general opinion among the experts who have been here is that a considerable amount of dynamite exploded in the open, which was exploded by or caused a blast in one or more holes in the heading, or was exploded by some other factor. This explosion probably took place on the lower left or west side of the tunnel at or very near the face, probably between the jumbo and the face, which is at Station 52+57. the men were loading the heading at the time of the explosion and the cause of the explosion is not known.

#2

A list of the dead and injured follows:

Willis Lee, 2212 Hunter St. Balto.} Dead

Chas. Durham, 2212 Hunter St. Balto.} Dead

Luther Randolf, 106 N. Bruce St. Balto.} Dead

Robert James, 811 Forrest St. Balto.} Dead

Melvin Dukes, 1413 Ward St. Balto.} Dead

Henry Sanford, 1515 E. Biddle St.} Dead

James Hough, 1510 N. Dallas St. Balto.} Dead

James Kay, 1012 Aisquith St. Balto.} Dead

Chas. Harvey, 1316 N. Mount St. Balto.} Dead

William Botts, 2556 Mace St. Balto.} Dead

John Reed, shift boss, 2736 Maryland Ave.
} Injured

Chas. Elliott, Electrician, 515 N. Linwood Ave.
} Injured

Christopher Tiggle, 1427 Argyle Ave.
} Injured

Matt Colston, 907 Valley St.

} Injured

Wm. Geary, 11 N. Chester St.

} Injured

All of the above men are employees of the Shea Company.

Phillip Lesser, 3800 Towanda Ave.

} Injured. Explosives Inspector, Bureau of Highways

As there are to be more investigations made the heading will be left in its present condition until we are allowed to go on by the various authorities.

Copies of reports made by Chief Inspector Henthorn, Inspector Baryle, Explosive Inspector Burns and verbal statements made by John Read and Christopher Tiggle to the U.S. and Md. Bureau of Mines Officials and myself are attached.

I have been busy with the various investigating committees for the past two days and will make a complete report of the activities later.

Yours truly,

John Hunt

Following is a copy of the statement given by Phillip Lesser to the City Solicitors Office, attested to by E.P. Roth on July 23, 1938:

I am employed by the City of Baltimore in the capacity of dynamite inspector and I have been assigned to the New Montebello Water Tunnel.

About 11pm on July 19, 1938, I reported for work at the office maintained by the City on Rost Ave., west of the Hillen Rd. This office is approximately 700 ft. from the Montebello Shaft. My shift on this particular night was from midnight until 8am of July 20. The work is so conducted that a charge of dynamite is exploded just before a shift goes off duty in order that the relieving shift might muck it out.

Approximately 6am, Foreman Reed telephoned that all drilling was finished and he was ready for powder in order that the holes might be charged. I walked over to the shaft, got Reed's order for eight boxes of dynamite and 61 primers and took the men assigned to me and went to the dynamite magazine to fill the order. The dynamite magazine is located approximately 300 ft. north of the Montebello Shaft and it is kept locked at all times and only dynamite inspectors are allowed to carry the keys. The primers are kept in a second magazine approximately 500 ft. north of the shaft.

The contractor provides a flat car upon which are mounted two wood lined steel boxes measuring about 5 ft. long, 2 ft. wide and 3 ft. deep. These boxes are large enough to hold nine boxes of dynamite.

After the dynamite was placed in one box and the primers in another, the car was lowered in the shaft and pulled by an electric locomotive to the heading. I, of course, accompanied the dynamite car and as soon as the dynamite was placed on the flat car, the trolley is pulled down on the locomotive and it is propelled from a battery. As we approached the heading I noticed that the drillers were just finishing up so the dynamite car was sidetracked on a California[34] siding about 100 ft. from the heading. Upon the completion of the drilling, the holes were blown out and then all loose steel was placed on the steel car and hauled toward the shaft. The shifter then gave the signal that all was clear and ready for loading. The electric current supplying light, trolley wire, pumps, etc. was then disconnected and the car was then moved up to the heading. Lights are supplied from a 250 candlepower light that operates from the electric storage battery and this light is mounted on the locomotive. Additional light is supplied by the electric flashlights of the men.

-2-

When the dynamite car reached within a few feet of the heading, the loaders removed the dynamite and carried it to their stations. Five boxes were removed in this manor, three of which were placed on the upper level of the jumbo and two on the bottom level. In this manner, the dynamite is in proper position

---

[34] California switch - Train track switch in tunnel.

for the loading crews as each of the four corners and the center are loaded simultaneously. The primers were then removed from their box and distributed to the five crews by the shifter. Each crew received ten primers. These primers are all prepared by a man by the name of Long who is an employee of the contractor and each primer is marked. The primers are kept in separate compartments in the box and are divided as follows, non-delays, first and second delays, all in the first compartment. The third and fourth are in a second compartment, the fifth, sixth in a third and the seventh, and eighth in the last compartment.

The dynamite has also been prepared by Long and this preparation consists in splitting the side of the heavy waxed paper in order that it will spread after it has been placed in position and tamped.

Foreman Reed who is the shifter reported to me that 49 holes had been drilled for this blast. Reed has a certificate and is a fully qualified person to handle and use dynamite. He is in full charge of the loading operation and gives all the orders. My duty is to see that all dynamite is accounted for and to see that no regulations are violated but I do not direct any of the work.

The loading of these holes began at 6:20am as I asked Mr. Reed the time just as we reached the heading. The first or cushion stick was placed in each hole and then the primer to which is attached 16 ft. of wire was put in next. This primer stick is put in extremely easy and is not tamped. A third stick is then put in and is gently shoved into place. After that, the sticks are loaded two at a time and tamped. The tamping is for the purpose of shoving the sticks back into proper position and to also open up the cut sticks in order that all air chambers will be filled up and thus avoid misfires. The drill holes are about 2 inches in diameter and the dynamite about 1-¼ inches so the cutting allows it to spread and completely fill the drill holes. The tamping is done with an oak pole from 11 to 15 ft. in length, which has a blunt nose. These poles are removed each time a charge has been set and approximately two inches sawed from the end in order to prevent any possibility of using a sharp pointed tamper.

During the loading operation, I stood behind the left side of the jumbo and about 13 or 14 ft. away from the heading, in order that I might watch the operation of each of the five crews. I saw each of the crews load one hole and my next recollection is when I recovered

-3-

consciousness to find the tunnel in darkness. I got to my feet and in trying to get my bearings, I walked into the stonewalls of the tunnel several times. In groping around, I stumbled over a man and assisted him to his feet and by this time, I recalled my electric flash light. I instructed this man to hold onto me, as I would lead him out. The smoke and gas were so thick that even with the aid of my flash light I could hardly see a thing. When we had gone a distance of several hundred feet, I saw a third man who was just clear of the smoke and he also accompanied us toward the shaft. Just before reaching this man who was leaning against an east rib, we saw Mr. Geary, who is the motorman on the electric locomotive and he helped us aboard. At the cage, which was at the bottom of the shaft, we saw an old colored man whom we call Minnie and Mr. Pritchard the track boss. On the way to the shaft, I saw someone at the telephone but this was about the time we picked up Christopher Tiggle, the colored man who was leaning up against the rib. The first man I helped was named Watts and he is also colored. When we got on the cage, it was taken to the surface. Thomas Evans the muck boss was on his way out with the steel car had already reached the surface and he had reported the premature explosion. Mr. [Bayle], a Water Department Inspector, took me to Union Memorial Hospital in his car. Tiggle was taken to the hospital in Mr. Daily's car by a man by the name of Carter and within a short time several municipal came to the shaft.

At the time of the explosion, I saw or heard nothing and my first knowledge that something was wrong was when I recovered consciousness. I did not see any violations of any regulations either in the handling or use of the dynamite and there were no lights of any kind near the heading with the exception of those mentioned above. I personally saw that the lights from the California to the face were disconnected. No wires were connected to the shooting wires and I have no explanation of what have caused the premature explosion.

The finished lines of the tunnel measure 14 by 14 ft. although this distance is sometimes exceeded when there is an over breakage.

The members of this crew were all good and careful workmen and were those that remained after the gradual weeding out that had taken place during the time the job had been in operation. I had never had any trouble with any member of the gang for violating any safety regulations and Mr. Reed, the shifter was always extremely careful and a very capable man.

I am unable to state just what holes were exploded, as I did not see the heading after the explosion took place.

In the second paragraph I stated that the work is so conducted that a charge is exploded just before going off duty, which

-4-

is not entirely so since the contractor, of course, drills and shoots as often as possible in order to make as rapid progress as possible. It oftimes happens however that a charge is set off just before a gang goes off duty.

Leon Small writes this lengthy report to Frank Duncan:

Mr. Frank K. Duncan

August 1, 1938

Chief Engineer,

Dear Sir:

The following is a report on the premature explosion of dynamite that occurred in the Montebello heading of the Gunpowder Falls-Montebello Tunnel of this Bureau on the morning of July 20, 1938, the construction of the improvement being a contract undertaken by the J.F. Shea Company, Inc., Los Angeles, Calif.

The excavation for the tunnel is being carried on from six headings, each on a twenty-four hour schedule made up of three 8-hour shifts per day for five days per week, starting Sundays at midnight. The last regular round in the Montebello heading was fired at 11:15pm on July 19, this round carrying the excavation to station 52+57, the 0+00 station being the centerline of the Montebello shaft, which is the terminal end of the tunnel. The shift of miners and their various helpers, under a foreman, that was on duty at the time of the explosion began work at midnight on Tuesday, July 19, and proceeded with their routine job of removing the muck pulled from the heading by the last round, setting up the drill carriage, or jumbo, and drilling the heading preparatory for the next shot, and were getting ready to load the holes, all to the end that in the normal course of events the heading would be shot shortly before the end of the shift, the work being scheduled and manned so as to have each shift accomplish a cycle of operations, thus getting three shots per day at each heading.

At about 6am on July 20 the foreman in charge of the operations at the heading telephoned to the surface that the drilling was completed and the crew was ready for the dynamite to charge the holes. This message was given to Phillip Lesser, 3800 Towanda Avenue, the City's dynamite inspector on duty at the time, this man being an employee of the Bureau of Highways and assigned to the tunnel job. Succeeding the receipt of the telephone message, Lesser went to the Montebello shaft at the surface, got the foreman's order for 8 boxes of dynamite and a quantity of primers, and took the men assigned to the duty of getting this material to the magazine to fill the order. The dynamite and the primers were drawn from separate magazines and placed in separate wood lined steel boxes placed on a flat car. The car was lowered in the shaft and pulled by an electric locomotive to the heading. Lesser accompanied the explosives and, according to his report, as the dynamite car approached the heading he noticed that the drillers were just finishing and so the dynamite car was side tracked about 100 feet from the heading until the drilling was completed, the holes were blown out, and all loose drill steel and other unnecessary material was loaded on a car and started toward the shaft. The signal was then given that all was clear and ready for loading. Electric current supplying lights, trolley wires, pumps, etc., were then disconnected and the dynamite was moved up to the heading. During the loading operation, all sources of electric light and power are completely removed from the heading and the necessary illumination is provided by special floodlights connected to a storage battery of the locomotive, with such additional

-2-

lighting as is required supplied by electric flashlights of the men.

Phillip Lesser was the only employee of the City present at the heading at the time of the accident, his duties of dynamite inspector being to see that all dynamite is accounted for and that no regulations set up for the safe handling of explosives are violated, but these duties do not include the direction of any of the work, this being the responsibility of the foreman in charge of the crew. The loading of a heading is normally done by five miners, each with an assistant, and this normal crew of ten men undertook the loading at 6:20am, Lesser stating that the foreman told Lesser this was the time when Lesser requested it of the foreman just as he reached the heading. The jumbo that is used for drilling is kept at the heading for the loading operation, as it affords a working platform that enables the holes in the upper part of a heading to be reached. The attached photograph, identified as 76*B, shows the front view of the Montebello jumbo after it was removed from the heading about a week following the accident, and in this may be noted the collapsed platform, which consists of two halves pivoted about a hinge arrangement so that when the platform is up it affords a substantial working level approximately 6 feet above the floor of the tunnel. On this platform, six men were starting to load the upper holes of the heading and beneath it, from the floor level; four men started the loading at the bottom holes. Some of the primers had been placed in their proper holes and several boxes of dynamite had been distributed to the men at the heading and loading of the holes was proceeded in a routine manner, Lesser's report stating that at this time he was standing behind the left side of the jumbo and about fifteen feet from the heading, from which point he could watch the operations of each of the five 2-men crews doing the loading. Lesser states it is his recollection that he saw each of the crews load one hole and then, without being aware of the flash or hearing a sound, he lost consciousness and recovered to find himself lying on the floor of the tunnel and the tunnel in total darkness. Lesser states also that the first knowledge he had that anything was wrong was when he recovered consciousness; he avers that he did not see any violation of any regulation in the handling of the dynamite and is positive that no lights of any kind were near the heading, with the exception of the usual flood lights mounted on and supplied by the locomotive. Lesser personally saw to it that the lights between a nearby switch and the heading face were disconnected, he is positive no wires were connected to the leads of any primer, and he has no explanation as to what might have caused the premature explosion.

Apparently, all ten men who were doing the loading directly at the heading were killed; there is a possibility that an eleventh man may have been at the heading, but this man was badly injured and an attempt to get a statement from him at the hospital has not been made up to the time of this report. A list of the dead and injured follows: [See list above, John Hunt report].

-3-

All of these men were employees of the Shea Company and all of them colored with the exception of the ones indicated by an asterisk [Reed, Elliott and Geary]. Lesser suffered certain injuries, the extent of which is not known to the writer, although these cannot be very serious, as he was sent to his home by the hospital to which he was first taken shortly after he arrived.

Succeeding the explosion, all possible aid was given to the stricken crew and in this work the Police

Department and the Ambulance Division of the Fire Department rendered excellent service, although unfortunately, the condition of most of these men was such that nothing could be done for them. The smoke from the explosion had barely been cleared from the tunnel before an investigation was started for the purpose of determining, if possible, the cause, and since the explosion there have been a number of inspections of the scene made by experts in the explosives field, these including representatives of the Maryland State Bureau of Mines, the U.S. Bureau of Mines, the State Industrial Accident Commission, office and field engineers of the Hercules Powder Company, an electrical expert sent on a special assignment by the U.S. Bureau of Mines, representatives of the labor organization under which the work is being done, employees of the Shea Company, and the engineers of this Bureau. The writer has seen no report from any of these investigating bodies, but it is his understanding that there has been no criticism on the part of anyone aimed at the means that are taken in the storing and handling of the explosives and the precautions that are observed in the loading and firing of the rounds.

The writer visited to Montebello shaft shortly after the last ambulance for moving the dead and injured was driven away, but at this time he did not go to the heading, as he was informed that a crew was then at the heading recovering loose dynamite and primers from the vicinity of the heading, it being further stated that nothing whatever at the heading would be disturbed until every opportunity had been given for a complete investigation. The recovery of the loose dynamite following the explosion accounted for all except about one-half box, approximately fifty sticks, and apparently, this is the quantity that was involved in the catastrophe. Early on, the morning of July 21 the writer went to the heading and critically examined everything that would have a bearing on the cause and results of the disaster. At this time the heading presented the view shown in the attached photograph 73-b[35], except the jumbo was then in place and there was more general litter on the floor of the tunnel at the heading. A careful scrutiny of the holes indicated to the writer that only one of these had fired, this being the one circled in ink on the photograph. This particular hole shows evidence of having shot only at the front end, or collar, this portion of the hole being funnel-shaped, with a maximum diameter of about 5" at the face, and tapering back for a distance of 4 or 5 inches to the diameter of the hole as it was drilled. Directly beneath this hole there can be noticed a pool of water in the floor of the tunnel and this is approximately the size and shape of a large dishpan and

---

[35] All mentioned plates in this report are missing from the archives and no photos have been found.

82

with a depth of about 20" at the center. It is the writer's conclusion that a portion of a box of dynamite exploded on the floor at this point and possibly this explosion set off a primer that had been partly entered in the hole that shows evidence of shooting in the collar; in the upper center of the heading may be noted a stick of dynamite with a

-4-

detonating cap inserted therein, this being one of the primers to which we have referred. This primer is being held by its wire leads that are stuck into the hole above it and no doubt, this primer was sticking in its hole at the time of the explosion, but was shaken loose to its present position by the blast. Had a similarly placed primer been in the hole that shows evidence of shooting, an explosion of a quantity of dynamite on the floor of the tunnel directly beneath it would very likely have set it off, with the results noted. No other holes in the heading show any apparent evidence of having shot, although the investigators from the U.S. bureau of Mines are inclined to think that possibly two other holes "gunned", and have taken scrapings from the sides of these holes for analysis.

Photograph 74-B shows a side-front view of the jumbo after it was removed from the heading and a particular point of interest about this picture is the manner in which the vertical front supporting leg of the jumbo was bent backward and, not evident in the photograph, had the bottom end of the leg torn off. This leg was directly in front of the hole in the floor mentioned in the comments on the photograph of the heading, and it consists of 4" extra heavy steel pipe, this having a wall thickness of a little more than 3/8 of an inch, a fact that is indicative of the forces of the explosion and practically exclusive of the assumption that the disaster was caused by an explosion not external to the heading, although it is possible that the main explosion resulted from a lighter one in a hole. In addition to bending and tearing off the jumbo leg, the explosion splintered in an upward direction the platform of the jumbo as shown in photograph 76-B, throwing the men on the platform violently against the roof of the tunnel, as the remains of their broken hard hats and the badly shattered condition of their legs indicated.

Photograph 75-B shows another side-front view of the jumbo, this being similar to the view shown in photograph 74-B. The supporting leg in this case is intact and is indicative of the condition of the damaged leg of the jumbo, shown in 74-B, prior to the explosion.

A coroner's inquest was held at the Northeastern Police Station on the evening of Friday, July 29, and it is the understanding of the writer that this hearing was completed, although up to the writing of this report the writer has been unable to learn anything regarding the verdict. The testified at this inquest and, upon examination, gave the following as four possibilities that may have caused the explosion:

(A)   - A workman on the platform of the jumbo may have accidentally kicked a loose tool from the platform toward the floor of the tunnel, this striking a primer with sufficient force to fire it, the primer, in turn, setting off part of a box of dynamite.

(B)   -One of the workman on the jumbo platform may have accidentally dropped a primer incident to removing it from a hole preparatory to starting the loading of the hole with one plain stick of dynamite, which precedes the primed.

(C)   -One of the workmen may have attempted to unwind the approximately 14 ft. of lead wire loosely wrapped around each primer by holding the loose end of the wires and whipping the primer toward the floor, the primer thereby striking some hard sharp

-5-

object and exploding. Incidentally, this practice is forbidden and men have

85

been warned that doing it would result in their instant dismissal, but instances of this nature are known to have occurred.

(D)   -A piece of rock may have scaled from the roof of the tunnel and fallen directly at the face, hitting a primer and causing it to explode. This is the least probable of the four possible causes, as loose rock is always scaled from the roof preparatory to drilling and the vibration of the drilling would very likely bring down any material that was overlooked.

The above outlines the circumstances surrounding the explosion and the results of the investigations that have been made by this Bureau and the writer to date. More technical information may be available when the various authorities that have investigated the explosion report their findings, but the writer is convinced that the real cause of the explosion will never be known.

Yours very truly,
(Signed) L.S.
Water Engineer
LS: B

News of this accident reaches out west to California and on July 31, 1938, Elizabeth Griswood of Oakland writes asking for information on one of the miners killed in the blast:

> ... name Robert James. Age make me think is my brother which I has not heard from in number years. Please send me his description height – color – weight. Also, if he was married please have his wife get in touch with me at once. I am enclosing self-address envelope and stamp for a reply.

Hunt sends off a short memo to Strohmeyer with the following information, on August 8:

> ...Robert James, Negro, killed in Montebello Heading ... age 36, 811 Forrest St., married, one child, wife Beulah James, description: black, 160#, 5'6".

Strohmeyer then forwards this information to Mrs. Griswood.

The reports, memos and letters above were included for the most part in their entirety, as submitted, and then filed in File Folder #1435.

Between August 2 and the 4th, there were various correspondence between the Explosives Inspectors and the Engineers concerning a stray stick of dynamite with a stone stuck in it, the loss of the shooting switch key, suspicious former employees on site and an anonymous phone call telling Werner to keep all city men out of the shaft. Burns writes to Hunt on August 2, 1938:

> A stick of dynamite was found some place on a motor ... the shunts were off of four primers. One loaded cartridge felt kind of hard. Upon further examination I found a stone about the size of a nickel imbedded in a stick of dynamite with a detonator ... After we had finished loading the heading and the extra dynamite was taken out I gave the key to the shooting switch to the shifter [McFadden], this he said he dropped. This key also fits the locks on the magazines. I could not find this key.

Hunt then follows up with a memo of his own to Strohmeyer on August 3, 1938:

> Four of the primers ... shunts broken ... extremely dangerous if they come in contact with electrical current ... The position of the stick of dynamite on the electric motor makes it almost a certainty that it was placed there and did not get there by falling.

Paul Sullivan, an explosives inspector, writes in his report to Mr. Hunt:

> One of the welders handed me a stick of dynamite and said it was found on the motor ... Mr. Shimmer, electrician ... informed me that there was a stick of dynamite lying in the ditch between the heading and the California switch.

> A letter from B.L. Werner, Assistant Construction Engineer, reiterates all of the above with this added P.S.: "At

88

Montebello tonight were Roy Graner and 'Blackie' Duroff, former employees of the J.F. Shea Co. I do not know whether or not this is significant." On August 4, Leon Small writes to Duncan:

> ... sending you copies of statements ... contain all information I have ... although Mr. Werner reports that at 11:00 pm August 3 he received a telephone call at the office of the Montebello shaft from a man who warned him to keep all city men out of the shaft. The man's voice seemed to be disguised and could not be recognized by Mr. Werner.

A PS at the bottom reads, 'Mr. Small, Mr. Duncan released this information to the newspapers Thursday, August 4, at 3:30 pm. J.S.S. 8-5-38'

The investigation into the tunnel explosion seemed to shift towards 'safe handling' of the explosives and whether or not proper procedures were followed for the guarding of those explosives. On August 30, 1938, Iardella types out a memo to Walter King and Anthony Topoleski, Special Agents of the P.W.A. answering their verbal request for information about the safe handling of dynamite (Ordinance No. 132) and the weekly pay for watchmen ($13.50). This information was then forwarded to the City Solicitors office. Then on September 3, Mr. King writes to Leon Small for clarification:

> Paragraph 1-12, 'Use of Explosives', page 38 of the specifications covering the construction of the Gunpowder Falls-Montebello Water tunnel contains the sentence:

All such storage places shall be marked clearly 'Dangerous Explosives', and shall be in care of competent watchmen at all times.

Information is requested as to the established rate per hour for watchmen,

Have watchmen been employed in accordance with this specification, if not, has the City waived this particular provision of the specification or approved the non-employment of watchmen;

Has the contractor ever made a request for a relaxation of the requirement?

Were any written instructions ever written by the Water Department as to the duties of the Explosives Inspectors on this project, if not, what oral instructions were issued to the explosives inspectors?

It is assumed that Mr. Small turned this request for information over to Mr. Hunt because on September 4, Mr. Hunt writes Mr. Small the following:

Subject: Statement regarding questions asked by Mr. Walter W. King, Special Agent, Division of Investigation P.W.A.

The established rate for watchmen is $13.50 per week for a 40-hour week.

No men have been employed for the specific purpose of watching the magazines --- that is, no man sat at the magazine and did nothing else. The City has not waived any provision of the contract insofar as watchmen are concerned. The working area of each shaft and portal has been fenced in, the dynamite stored in well built magazines, which are always locked. When the project started and up until recently the men worked six (6) days a week and maintenance work was done on Sunday. Recently the men went on a five-day week, maintenance work being done on Saturday and Sunday. This means that there were always men on top in the compressor room and as a usual thing, there was always maintenance men working on Saturdays and Sundays, when the job was shut down. It has never been the practice of the City to require watchmen at magazines who did nothing else. The ordinary watchman, pump man, or other maintenance man has been considered all that was necessary, and the fact that the work was fenced in, magazines locked, and men on duty inside the fence at all times was sufficient to comply with the specifications.

Mr. Hunt goes on and states that the contractor never requested guards for the job and there had never been any written instructions for the explosives inspectors on how to do their job. As these inspectors were on loan from the Bureau of Highways, they followed their own protocol for the safe handling of explosives. It was the responsibility of the water tunnel inspectors to make sure the explosives inspectors were doing their jobs, and in a safe manner. There were oral instructions given by the Construction Engineer that covered the basics, such as:

No dynamite is to be taken out of the powder magazine unless the Explosives Inspector was present. The Explosives Inspector to see that the primers and powder be put in separate boxes and taken to the heading in a cautious manner, --- safety and firing switches in proper position. The electric lighting system in the heading be pulled back a safe distance before loading is started. The heading loaded in a safe manner, wires hooked up properly and all men out of heading, safety switch put in firing position, and blast made in presence of inspector. The inspector made reports giving the time of blast, number of holes fired, number of primers used, amount of power, etc. All other safety rules, such as not allowing smoking while loading, no welding with open flames, rough handling of primers, hard tamping, etc., to be strictly enforced. Misfires to be watched for and handled in a safe manner.

He concludes his memo with this brief summary: "The specifications were not changed or modified as to the safe handling of dynamite on the job." This memo is then forwarded to Mr. King. Also on September 4, Mr. Small sends off another memo, this time to John Horan, Special Agent in Charge, Division of Investigation for the P.W.A., requesting copies of all statements made by City employees concerning the tunnel blast. Ten days later, Mr. Horan replies, "It is not permitted to make available to you copies ..." but suggests that Small ask the men interviewed for a copy, as copies were given to them. On the 16th, Small writes back that he should be allowed to have copies, not so much concerning the explosion, but those copies from employees that made comments concerning specifications for the handling of dynamite. Horan appears to give in to Small for the copies and asks for the names of the employees from which he would like copies. On the 23rd, Small sends the list: Werner, Bayrle, Keim, Sullivan, Burns, Wachter, Lesser and Isaac. This request is then forwarded to the Acting Special Agent in Charge, Joseph Bishop, and he replies to both Horan and Small:

> In view of the confidential nature of the investigation and the fact that the report, of which the statements requested are component parts, is confidential and not for public inspection, I have referred your request to the Director.

Horan writes to Small on September 30, 1938 that "… the records are confidential and not for public distribution."

When the investigation finally winds down, Bayrle writes in his journal concerning other personnel matters and tunnel progress:

> September 2 – Lehnert was two hrs late tonight and he did not call up or offer any excuse whatever when Werner asked him 'how come?' He cursed Werner and wanted to fight him, said Werner had no authority over him. This man should be fired. He also threatened Werner, if Werner reported him. Haw – haw! Due to Lehnert being late, Sullivan had to examine the powder and primers alone, make the shot and then go back in

the hole after the shot was made. Lehnert again made no effort to relieve Sullivan. Mr. Strohmeyer made a ruling today – that city inspectors were not to sign any statement or [talk to] any investigators unless ordered to do so by the City Solicitors Office. September 13 – All powder men except Keim sent into office today. Keim is on day shift until further notice. No blasting. The powder and cap houses were cleaned out today and the detail of detectives were taken off the Montebello shaft. The grade party checked the tunnel between Louise and Montebello today. Centerline checks ¼" and grade checks 0.04'. This is perfect work.

Chapter VIII

Work resumes, Dry wells and noise

For the rest of 1938, it was business as usual at the tunnel. There were some tests done on the residents' springs and wells around the centerline of the tunnel, as well as test for vibration. The majority of complaints concerning the wells drying up came from the residents living in the Westmoreland Ave. area. It became of such concern that both the residents, through a lawyer, and the City, each hired a geologist to investigate the matter. Geologist Joseph T. Singewald, for the City, notes in his investigation that:

> In 1936, before the driving of the tunnel had commenced, your Bureau measured the depth of water in a number of these wells at intervals extending from late in winter to late in the following fall. [Table illustrates a maximum 6.4 feet to a minimum .1 foot in 1936, to being dry in four out of five wells in 1938]. They (the data) demonstrate that in 1936, before the driving of the tunnel had commenced, the ground water level had receded due to entirely natural causes ... A comparison of the precipitation ... 1935-1936

rainfall: 31.12 inches, snowfall: 30.0. 1937-1938 rainfall: 24.28 and snowfall 8.9 ... the snowfall in 1938 was much more unfavorable than it was in 1936. Melting snow is much more effective in increasing the underground water supply than is an equivalent amount of rain ... The tunnel in the Taylor Ave. area lies about 190 to 220 feet below the surface. The intervening rock is the Gunpowder Granite. This is an extremely dense rock that could not serve to drain the underground water supply into the tunnel ... Natural causes alone are adequate to explain the condition of these wells. The tunnel may have been an additional factor.

Mr. Singewald makes note in a second memo of the same date (Nov. 15) that he has been contacted by Dr. Harry Fielding Reid, Emeritus Professor of Dynamic Geology, at The Johns Hopkins University and that he (Reid) has been retained by a lawyer "... to look over the situation on behalf of all the complainants ..." Singewald continues:

> Yesterday Dr. Reid came to see me and said he would regret a situation in which he and I might be called on to testify in court and give contradictory testimony. He thought that if we conferred on the situation we could come to a mutual agreement, yet he said that since he had been retained by the other side he did not feel at liberty to divulge at this time the reason for his conclusion. He said his conclusion was that the tunnel was responsible for the wells going dry ... I told him that I regarded this an unproved conclusion ... I then

asked him if he thought the effect on the wells would continue after the tunnel was sealed. He answered in the negative. I then asked him if you [the City] continued to supply these people with water until the tunnel was finished, whether you would not have remedied any injury the tunnel could have done to them. He considered this a fair and adequate solution to the problem ... Our opinions differ with respect to the cause of the temporary situation ... In view of Dr. Reid's reluctance to state the details of his evidence, I did not feel at liberty to tell him the details of the 1936 well measurements.

Some of the neighbors complained about how their homes shook whenever there was an explosion. When the City did nothing about the complaints, the homeowners wrote to Mr. Fixit of the News Post[36]. In his journal, Watt explained the vibration test used in determining if the work going on in the tunnel was the cause of damage to homes:

Four, forty penny nails were set on [their] heads at locations shown on sketch. [They] remained in a stationary and upright position during blast. The position in which these nails were placed indicated there was little if any surface vibrations during blast.

These tests did not stop many of the homeowners from filing lawsuits against the City and the Shea Co.

---

[36] Mr. Fixit. A column in the News Post newspaper used to help citizens with problems, get results.

When some of the residents figured that the threat of a lawsuit was not enough, they took matters in their own hands, making harassing calls to Mr. Small at all hours of the night. In a March 9, 1938, letter from Small to the Telephone Company:

> Residents have gotten into the habit of calling the writer at his home around midnight, in order to inform him that dynamiting in the tunnel has disturbed their rest. A telephone call at this hour has a certain nuisance value which, no doubt, appeals to the parties making the calls, and in order to minimize such calls, it is our thought that your operators on the night shift may attempt to divert such calls to the Montebello office of this bureau, University 7693, where our engineers are on duty twenty-four hours per day.

Mrs. J. Logan Jenkins, of College Ave. in Morgan Park, told Kavanagh and Werner that "… all the test were faked." In her complaint of May 2, 1938, she tells them that the blast was so bad that a cut glass dish on the table had broken by a falling piece of chandelier. During a blast on May 3, a pin test was conducted. Inspector Rodgers writes to Mr. Hunt, "No motion of pins was observed but shot sounded very loud." The shot consisted of eight cases of dynamite with forty-eight primers. Another complaint was made on May 4, where Mrs. Jenkins shows Werner a lampshade that fell over, breaking a string of beads. Another pin test performed that afternoon and the following day, both of which had negative results. On May 6, another test is set up but this time "A mop handle standing against the Newell post in hall

98

slid to floor but I had difficulty making it stand up again due to slippery floor." By August 4, Mr. Small writes to Mrs. Jenkins stating that the Bureau disclaims all responsibility.

Another complaint of the neighbors concerned the noise made by the diesel engines at the portals and shafts. One complaint, by a Mr. September Dime[37], who lived 1-1/2 miles from the Miller shaft, claimed, "… the noise from the diesel engines … are slowly driving him insane and he cannot tolerate it much longer." Werner, who visited his home, said he could not hear a thing from inside the house. Mr. Dime stated that he would like the City or the contractor to buy him a new house. Werner concluded his report with this observation: "It is my opinion that Mr. Dime is suffering from a nervous disorder." After a visit by Mr. Dime, Leon Small writes a memo to Mr. Kavanagh:

> I feel convinced there is insufficient noise to annoy any
> normal person residing even closer that does Mr. Dime
> and I therefore think that his condition is abnormal;
> but this very abnormality, in my opinion, is a menace,
> as this man may do something desperate.

Mr. Kavanagh has the same opinion, that Mr. Dime is not a well person and his opinioned insanity, is about to be tested. A memo by Strohmeyer:

> I called Mr. Kavanagh regarding Mr. Dime, who visited
> me again today and made a long complaint about the
> noise of the diesel engine at Miller shaft. Mr. Dime

---

[37] Name has been changed. Relatives still live in this area.

stated he had just found out that there is no muffler on the engine at Miller as there is at Louise and Montebello shafts. I related this to Mr. Kavanagh and he stated that he intended to make a test tonight by shutting down the diesel at Miller and calling Mr. Dime and talk to him to determine if he still thinks he hears a noise; and as he has a complex that we are trying to deceive him, in all probability his answer will be that he does not hear it, and as he has no telephone one of our men and one of Kavanagh's men will call again later at night and see if the noise can be heard, this being apparently conclusive if Mr. Dime's delusion is caused by an illusion of his own mind, or if he does really hear something caused by the hypersensitivity of his hearing.

This test took place and at about 8pm, with the engine off for 45 minutes, they visited the Dime's and both the Mr. and Mrs. said they could hear the engine. At 9am the next morning, they paid another visit and again Mr. Dime said they could still hear the noise "As about as bad as any other night and that it was useless for them to be bothering them by calling on their home."

Another visit, three months later, by Mr. Dime to Mr. Strohmeyer, is noted. In a report to Mr. Small, Strohmeyer states, "Mr. Dime visited me in a highly nervous condition … if the noise is not stopped, he will stop it himself." A month later Mr. Dime writes to Mr. Fixit at the Baltimore News Post, who investigates his claim. After talking to the engineers and inspectors, they are reassured everything possible has been done to appease Mr. Dime.

A month later, the project is ending and there are no other complaints recorded from Dime.

The tunnel was built in sections, from one portal to the next. On August 5, the workers holed through between the Loch Raven portal and the Miller shaft. On September 9, the crew stopped working in the south heading of the Louise section and the next day, the Montebello crew holed through. Bayrle noted that the centerline and grade looks 'ok'. On September 13 most of the powder men are let go, except Keim and removal of excavating equipment begins. Bayrle's entry for that day:

> All powder men except Keim sent into office today. Keim is on day shift until further notice. No blasting. Removing blower pipe from tunnel. Pipe being stacked at Montebello yard. The big Ingersoll Rand Hor (horizontal) compressor is being torn out preparatory to shipping it to the Shea Co. job in New York. The powder and cap houses were cleaned out today and the detail of detectives were taken off the Montebello shaft. The grade party checked the tunnel between Louise and Montebello today. Center line checks ¼" and grade checks 0.04'. This is perfect work.

Henthorn mixes some personal notes and opinions in his work journal from December 5, through the 19th:

> December 5 thru 17 on vacation in the land of sunshine and roses. December 16 – Tunnel 95% complete. Not much for me left to do? What in the world have they done – T. December 19 – Tunnel

completed the day my vacation ended - ? Some Baloney
– What a job – Right where it was when I left. Will
have to be started* again.

*Note: The word 'started' was scribbled over the word
'rebuilt'.

For parts of November thru the end of December 1938,
the journal entries covered the beginning of the concrete work
and the problems of cold weather pouring. Below is a picture of
how the curbing along the bottom of the tunnel looked.

(Baltimore City Archives Lantern Slide #91)

Chapter IX

Troubles with construction

There were numerous entries made on the pouring and condition of the concrete. Both Henthorn and Curtis wrote about the problems with an even mixing of the cement, also how the weather conditions affected the pours. On November 16 and 17, 1938, Henthorn made a couple crude sketches along with these journal entries:

> Experimenting with attachable and adjustable chutes on truck chutes to prevent segregation in the car. Condition is thus: [diagram in journal shows cement truck, chute, gravel, grout, car, segregation]. As result of above, the curbing on the east side of tunnel is getting more of the grout than the west side. Go with Mr. Hunt to Kavanagh at Louise and have an attachable chute made to hang on the end of the truck chute so as to get a turn over of the concrete during its flow from the truck to the car. Try several types but none were as perfect as desired. Mr. Hunt suggested an adjustable one. Mr. Kavanagh says he will make one during the night. Mr. Hunt finds an article in the

Engineers News Record describing exact condition we have and correction for same. Go to portal and try out chute made thus: [diagram in journal illustrated below]. Chute works perfectly. Very little segregation in car. Mr. Hunt suggests that a rod be attached to the chute as shown by the dot and dash line so that a man standing on the unloading platform can move it at any angle.

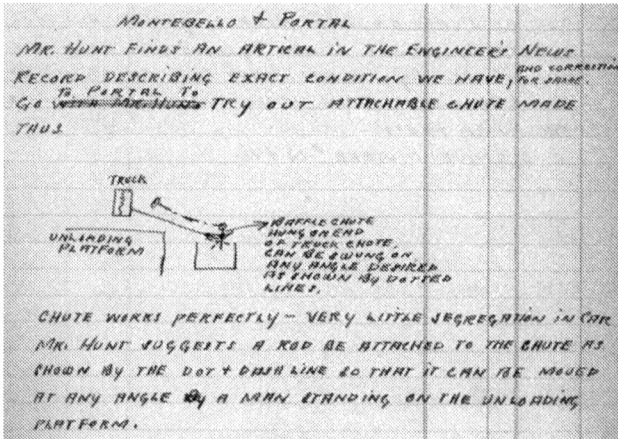

(Henthorn Journal, 1938)

Curtis kept track of the temperature and the delivery of the concrete:

Campbell plant [White Marsh] shut down. Too cold. Temperature standing at zero at 7:30am. Pouring curb. Temperature of water used in concrete = 118°. Sand and gravel coming from the plant heated.

The first nine months of 1939 covered the concrete work in the tunnel. Bayrle writes for January 4:

Working at batch plant, at Miller shaft today. Watt and Henthorn on the mixer and pouring bottom in the tunnel. Dost checking time for Miller and the portal crews. Cement for regular batches was sent in on car and added at the mixer. Having trouble getting Pump-Crete machine[38] working right.

On January 16, Watt makes a complaint about the quality of work and of the workers themselves:

Pouring concrete at invert - station 329+79. 171 linear ft. Batch stuck in chute at Miller delayed starting. Clean up gang loses too much time, no leader. Walked up and down the tunnel for 3 hours and when concrete caught up to end of clean section, the foreman wanted to pour over dry muck. Stopped concrete and placed bulkhead. Kavanaugh in hole 3:30 talking to Rood. Elmer Johnson's gang laying track and cleaning up.

Mr. Watt once again has to assert his authority as demonstrated in this journal entry for April 12, 1939:

One piece of shooting line cut off and rolled down wide place on slope. Delays off and on until 4am. Shut down 4am to reline and weld air booster connection to shooting line also broke four shear pins on pump. I think the valves on pump are out of adjustment. Sam Ross in tunnel about 3am and told me he was going to run concrete wet. I told him he would run the concrete the way I wanted it or the job would be shut down. We had one hell of a battle - I won.

---

[38] Pump Crete machine: Used to transfer concrete mix.

As Chief inspector, George Henthorn was delegated the task to oversee the manufacturing of the steel pipe that would be used to line the tunnel. He would also take welding classes and test the newly hired welders. His journal entry for June 8:

Note here from Hunt to be at Leetsdale (Bethlehem Steel Co. in Pennsylvania) Monday. Werner advises Rogers is to be taken to Leetsdale and broken in. At Portal with Williams. Had another argument with Williams. This time over the consistency of the concrete and his refusal to do as I wanted done, so I left the tunnel at 1am and the job to him. We were finishing the last pour in the extra work section. Phoned Werner and saw Mr. Hunt. Hunt advised he would send Williams to Montebello and I stay at portal tomorrow. 6pm – Hunt phoned house saying there wouldn't be any pour at Montebello tomorrow so for me to come in at 8am and see Blackburn about Leetsdale."

Two years later, the problem of dust in the tunnel is still not resolved as noted by Watt once again on June 27, "Made report to Hunt about cement dust in tunnel. Dust so damn thick, you could not see a man 16ft away." This was followed by another entry from Watt two weeks later, "Cement dust heavy in tunnel. My mouth, throat, and lungs have been sore for some time now because the contractor has not been made to correct this condition."

There were only two notes concerning the unions for 1939. One from Watt on January 25: "Still tells me the union can't furnish cement finishers and Carney tells me the contractor doesn't want finishers but wants to use men already on the payroll." and another from Henthorn on August 12: "Job shut down on account union trouble." Note: Could not find any reference to type of trouble in the journals or memos and letters.

Through out the 1939 journals were scattered bits and pieces of personal matters, both in and out of the tunnel. On a personnel issue, Watt made this journal entry on February 21:

> Fischer asks me if I had anything against him. I told him that I had nothing against him but he had to do his work like any one else. Fischer has gotten so damn big headed that he resents being told to do his work and wants his men to think he is a tin god and expects his men to disregard any instructions given by the inspectors.

Bayrle made this unusual personal note on March 9: "I did not work today. Ruth sick – measles." Followed by this one on June 22 concerning a favorite activity of the workers, "Big ball game today at Montebello. Time 4pm to 9. Tunnel men, water dept. vs. Office men, water dept. Stake – 1 keg beer (1/2 barrel)." Note: Could not find a reference as to who won.

In a July 24, 1939 letter From Rosa Garnand to the Chief Engineer, Baltimore City Water Department, Rosa expresses her concern for one of her neighbors:

Dear Sir, I would like you to know about Louis Stein who works on the City Filtering Station on Harford Road and always comes home drunk and abuses his good wife and children so they have to run to a neighbor's house for safety. She has heart trouble and many times has been in bed without coal to make a fire to keep warm or money to pay gas bill, coal or doctor or medicine. I went to see her Sunday and because he said she did not have the radio fixed, he kicked her in her side after which she was taken to her neighbors until he fell asleep. She is the mother of four children by him her first and only husband. He gives her 16 dollars a week and keeps the rest for himself. Now Mr. Chief Engineer, I am her sister and have and do help her out by sending coal or paying grocery bills and I wish you would investigate this case and see about this terrible drunkenness and abuse to the family. Hoping to hear from you soon.

Whether the Chief Engineer did anything about this is unknown.

Another personnel issue was between Watt and a fellow named Ballew. Watt writes on September 22, "Ballew cut down half a pine tree at Sta 263+00 which was about 25 ft off center line of trench and covered the stump with dirt. I gave him hell and told him if another tree was cut down with out permission, he would be fired."

On a somber note, Watt made this entry: "September 23 – Miller's store burnt out this morning about 9:30am. Children

108

upset kerosene stove in living room, flame from upset stove caught kerosene on linoleum from which flame spread rapidly." He followed his journal entry with a memo to Mr. Hunt, Watt describes the fire as such; "Flames spread so fast it was impossible for the Miller's to save any of their money or personal belonging. Wood in old building burned like dry hay."

Chapter X

Job extension and steel pipe woes

On March 30, 1939, Kavanaugh sent a memo to Mr. Small requesting an extension of five months to the contract, noting for his reasons the events that were not in his control, such as the labor strike and tunnel explosion. Mr. Small replies that the Shea Co. needs to make this request at the end of the contract, December 31, 1939 (Which they do on December 19, 1939)[39]. Meanwhile, there is a memo from the Federal Emergency Administration of Public Works, Mr. Shryock, to Leon Small pointing out the slow progress of the Shea Co. and asking Small to "Exercise your good offices in having him do actual construction work in as many points as possible so that his rate of progress will be accelerated." Small forwards this letter, along with one of his own to Kavanaugh, "You will note from Mr. Shryock's letter that he is perturbed about the slow progress ... desires that the rate of

---

[39] Chief Engineer Duncan writes the P.W.A, noting that out of 21 federally funded projects, only two are incomplete – the Montebello tunnel and the Curtis Bay (Patapsco) sewerage plant. The City had asked for no extra money and spent over one million of its own funds on the sewerage plant and the Municipal Airport.

110

progress be accelerated." Kavanaugh responds to Small that they are doing the best they can, noting other reasons for the slower progress. (In September, Kavanaugh asks for a nine-month extension because this portion of the construction, at the Montebello section, is "of a coordination nature").

By late September 1939, the PWA gets involved again with the progress of the tunnel and they ask Chief Engineer Duncan to submit a formal request for an extension. Duncan turns this task over to Small who writes a three-page document on why an extension should be granted. Small notes that the original contract was for 3-1/2 years but was reduced to two years and ten months, "a fact that indicates our judgment in the matter was influenced more by optimism then by reason". He then goes on and list other reasons for an extension, including that it took over a month just to give the contractor a "Notice to Proceed" document.

Then, once the contractor was on board, they needed to look at the problem of power for the various shaft locations and for the tunnel work itself. While negotiations with the local power company where taking place, orders for machinery with alternating current motors were being placed, but when negotiations failed and the contractor decided to build his own generating plant, the machinery needed to be modified for direct current, as supplied by diesel engines.

The large labor turnover at the beginning of the contract was also listed by Small. The reclassification of employees and the labor negotiations for the 40-hour week created a considerable loss of time, not to mention the tunnel explosion, which lessened the crew by ten workers and brought morale down. It was also noted that due to the various labor organizations fighting among themselves and going on strike was a cause for some of the delays.

One final note that Small makes, concerning the delays, was the result of design and construction changes to the lining of the tunnel: the contractor wishing to use the monolithic, one large piece, approach but the engineers opting for a sectional operation.

The Federal Works Agency in Washington D.C., which oversees all the construction and funding for the many municipal projects, shows their frustration at all the 'request for extension' and expresses that in a memo to Mayor Jackson on February 27, 1940. John Carmody, Administrator, had this to say about the latest request for an extension:

> It occurs to me that you may wish to talk to Mr. Cobb and perhaps other members of the engineering staff about this request. I doubt whether you are aware of the fact that as the record stands the delays already encountered put all of us in a bad light.

Carmody then goes on to list all of the requested extensions and reminds the Mayor that the last extension to June 29, 1940 stipulated "... the project be closed out on that date in so far as the P.W.A. was concerned and the applicant will assume financial responsibility." As far as the reasons given by the contractor for the extensions, other than the explosion of 1938, Carmody writes, all others should have been 'foreseen'.

Carmody continues:

There has been exceptionally poor planning throughout the construction of this project and lack of sufficient force of assistants in coordinating and completing plans, specifications, and other preliminary work. Frankly, these delays are indefensible for any kind of project anywhere. If it were generally known that we were approving another delay of several months, the hundreds of people who have reorganized their work in recent months to get done on time would think they had been discriminated against. I am sure neither you nor your engineers would like to have us give that impression now.

Mayor Jackson sends a brief, to the point, memo to Mr. Cobb – "PLEASE TALK WITH ME". Whether he did is unknown. There is, however, a four-page memo from Cobb to the Mayor responding to the Carmody accusations. He feels that all the P.W.A. jobs have gone well and that the tunnel project was underestimated from the beginning, but that is no reason for the P.W.A. or the City to feel embarrassed over the amount of time it has taken. On March 18, 1940, the P.W.A. sends a memo to the Mayor stating, "Consequently, this work, as well as all other work done on this project after June 29, 1940, must be done wholly at the City's expense …"

Meanwhile, both the Senate and House of Representatives have introduced joint resolutions (H.J. Res. 438), which were referred to the Committee on Appropriations. This resolution states: 'Extending time for construction of work relief and public works projects until January 1, 1941'. In addition, in the Baltimore Morning Sun newspaper, dated May 22, 1940, is an article entitled *Wants Time to Complete PWA Projects Extended,* Carmody was quoted asking congress to extend the deadline for projects to June 30, 1941.

On May 24, 1940, P.W.A. Administrative Order No. 233 (Supplement 3) was adopted, allowing the P.W.A. to allow extensions of time for projects. So on June 24, 1940, Mr. Cobb writes an eight-page request to extend the completion time of the tunnel to December 31, 1940. His request is denied and in a memo from the Commissioner of Public Works, E.W. Clark, of the PWA, states, "Notwithstanding the fact that design and construction difficulties gave rise to problems which required extensive experimentation, we believe that much of this could have been avoided and the project completed well within the construction period allowed." Mr. Strohmeyer takes exception to the 'experimental' phrase, and suggests in a memo to Mr. Small, "… this is an innuendo, as I do not otherwise see its application." The project must be done by September 30, 1940.

So the work continues and in October of 1939, the first section of pipe for the tunnel was lowered down the shaft.

(Baltimore City Archives Lantern Slide #64)

On October 7, Henthorn made this entry on the pipe that was to be delivered to the worksite:

> Bethlehem: George Transfer unloading pipe from car to truck in P.R.R.[40] yard. First one lifted from car and swung around when crane turned over and dropped boom on pipe, crushing one end of pipe.

Throughout the rest of the year, the inspectors were delegated the task of testing new welders. With the classes that Henthorn had taken previously, he was able to train the other inspectors. Peregoy, a new inspector, made this note: "October 17 – Testing welders. The first length of 12' 4" pipe was lowered into

---

[40] P.R.R, Pennsylvania Rail Road.

the hole at 2pm. Not much progress was made in transporting it up tunnel."

This was followed up by Henthorn's entry on October 18:

> Bayrle, Cherry, and Peregoy testing welders. Received one 12'4" pipe (total 14). Placing carriages in 2nd pipe. Conveying first pipe up tunnel. Had trouble getting through account track out of alignment and not level and not enough endplay in bearing guides on carriages. Kane advises Kavanaugh will have to correct track alignment. Tested four welders – 1 accepted, three rejected. Mr. Hunt advises to take charge of work going on, on this end of job with Bayrle, Cherry, and Peregoy. Will let us have another man if necessary.

The next day, they received a second section of pipe which was taken in 2300 ft. Shea's men took the pipe the rest of the way to the Louise shaft, testing the track as they went.

November 1939 was the beginning of trouble with the steel pipe. After setting the pipe and then grouting behind it, the crew found hollowing and then bulging occurred. Henthorn writes, "November 1 – Tapped pipe for solidity of grout between station 112+35 and station 110+95 – found hollow sounds from one end to the other. Advised Mr. Hunt. Going to cut into pipe to see what is causing it." On November 16, the problem got worse. Henthorn again:

> Bulkhead at 103+35 broke. Pumped 432 bags – 42.40 yards, when pipe bulged for a distance of about 50' and out 12" at greatest depth – width of bulge about 2 ½'.

Bulge was in upper east quadrant. Shea washing grout out from bulged section of pipe. Cut five – 5"x5" holes through pipe to get grout out.

(Baltimore City Archives Lantern Slide #103)

Besides over-grouting on the exterior of the pipe being a problem, water from underground streams started to bulge the pipe inward at Section "A" of the tunnel. This problem was alleviated by use of check valves throughout the tunnel length (about 350 check valves). When the tunnel was full, the check valves would close and when the tunnel was empty, the valves would open to relieve external pressure. In April of 1940, Mr. Small had to write the Public Improvement Commission to ask for additional funds, about $18,000.00, to pay the Shea Co. for all the work they were doing to correct the bulging.

Henthorn concludes his 1939 journal with these words of wit: "December 31 – End of another year and one year closer to St. Peter."

Chapter XI

Job completed, Mayor Jackson fills the tunnel

Henthorn's journal was the only one found for the year 1940. This journal gives brief summaries of the three different shifts placing pipe and grouting in tunnel. It also concludes with a final inspection and putting the tunnel in service.

The cold weather was again affecting the progress of the tunnel work. Remarked on January 2, that no pipe could be set and no pipe could be brought in because of ice on the street Also, no hauling could be done. Shea's men could only do some mucking in the by pass. They also stopped working the 4-12 shifts. Because of six inches of snow on the ground on January 24, the crew did not come to work. The grouting machine was also broke down. The freezing weather continued into February. Henthorn writes on the 8th:

> Removing scaffold out of by pass. Placed oil burner
> and wood fires in by pass to melt ice. Tested welder
> Moore of J.F. Shea for qualification to work on pipe.
> Rejected him on operation. Grout crew and welders

knocked off due to smoke in tunnel from fires in by pass.

Since the tunnel explosion, there had been very few entries on injuries and accidents in the journals. Inspector Werner did make this notation in a memo to Mr. Hunt on March 23, 1940:

> At 11 am today, I gave the signal at the bottom of the Louise shaft for the cage. Just after giving the signal, I noticed some cement lumps dropping down shaft into the sump. I was about to look up to see what was happening when there was a very load noise, like an explosion, my hat blew up the tunnel about 50 feet. When the dust cleared up, I saw the cage, very much twisted, lying at the bottom of the sump.

Apparently, some workers up top had sliced the cable while loading equipment.

A consideration discussed, but not really acted upon until the near completion of the tunnel, was the fact that dewatering equipment would be needed. In a memo to the Public Improvement Commission, Mr. Small explains the situation:

> The terminal end of the [Tunnel] is located at such a depressed elevation with relation to existing structures which will permit of gravity, drainage of the tunnel, that about 20 million gallons of water must be removed by pumping ... piping connections ... intimately related to tunnel construction ... included in tunnel contract ... but operating elements ... pump, motor, control switchboard ... were not included ...would have unnecessarily delayed the advertising of this contract.

119

The unwatering equipment must be installed and ready for service before any water is permitted to enter the new tunnel from Loch Raven.

In conclusion, to his memo, Mr. Small asks the P.I.C. for $48,000.00 and permission to advertise for this portion of the project. Permission was granted and after receiving and opening the bids on May 29, 1940, the recommendation was to go with the Dravo Corporation at a low bid of $42, 545.00.

Henthorn's journal continues with another weather related entry on May 20:

Heavy storm beginning about 4:30pm and lasting about 1 hour filled control vault with about 4' of water. Water from Tiffany Run backed up through the old tunnel into South valve vault then through the by pass and down the Montebello shaft. Electric pumps at bottom of shaft were fouled and the water rose to about 8' deep in shafts. Mechanics and crews working to get water out.

On September 21, 1940, as the work winds down, Edwin Peregoy is one of the first inspectors let go due to the lack of work. The City Service Commission is notified of his being laid off.

In a ceremony, fit for a mayor, on December 11, 1940 at 3:15pm, Mayor Jackson turned water into the tunnel at Loch Raven. This was not so much as to put the tunnel in service; the water was needed to do testing on the tunnel and dewatering station. Henthorn:

December 17 – Mr. Small out – trying pumps out. Test run until about noon. Mr. Hunt has figures. Started to drain tunnel about 5pm. Opened by-pass 60", valve three. The 4" eductor does not seem to be working – water comes back in shaft. About sixty feet of water in shaft. This excess water was pumped out.

December 21 – 6am: Start in tunnel at Montebello for final inspection with Laurie Leedum (Construction engineer for Newark, New Jersey), Messer Small, Blackburn, Strohmeyer, Hunt, Ningard, and Guman. Tunnel in good shape. 9:15am, Out at Loch Raven. Note: It took them three hours and fifteen minutes to walk the complete length of the tunnel, starting at the Montebello shaft and ending at Loch Raven.

December 23 – New tunnel put into service. In bottom of dry shaft found cement bags, rags, wire, water logged wedges and lumber, muck, etc. in the 4" eductor sump to a height of floor level. If the eductor had worked, it would have been a miracle. Took the eductor out and found it was choked in throat with wood chips and small pebbles. Sent it to top and it was thoroughly cleaned.

The tunnel was placed in service on December 23, 1940 but the entire project was not finished until April 18, 1941. The Gunpowder Falls – Montebello Water Tunnel, Docket No. MD. 1008-5-R, was formally completed and accepted on May 7, 1941 after members of the Public Improvement Commission conducted a final inspection of the tunnel. Final acceptance of the Gunpowder Falls-Montebello Tunnel is sent to motion, seconded and approved by the Public Improvement Commission on that date.

Nearly seventy years later, this tunnel is still providing fresh drinking water to Baltimore City and surrounding counties. For many years after the tunnel was built, walk-through inspections were conducted to check on it's condition and it was found to be in as good as shape as the day Mayor Jackson first turned a valve to fill it.

www.ingramcontent.com/pod-product-compliance
Lightning Source LLC
LaVergne TN
LVHW090046090426
835511LV00031B/331